FROM CHALK TO CHATBOTS

FROM CHALK TO CHATBOTS

Editing, Design & Layout by: Lena Mirembe Pira
Leap Book Publishers Limited (U)
P.O. Box 71990, Kampala, Uganda
info@leappublishers.com | +256-200-939350

Printing by: Joel Billionaire
Honey Badger Limited (U)
Nkrumah Road, Kampala, Uganda
honeybadgerprintery@gmail.com | +256-783-097947

Digital eBook design and formatting by:
AfricanBooks.com

First Edition [American English]
Printed in Uganda.

ISBN **#978-1-959070-21-4** (Softcover)
ISBN **#978-1-959070-29-0** (Hardcover)
ISBN **#978-1-959070-22-1** (eBook)

FROM CHALK TO CHATBOTS

DISCOVERING AUTHENTIC EDUCATION WITH AI

LAWRENCE MUGANGA, PHD

CONTENTS

INTRODUCTION

In a world rapidly advancing with technology, there is a silent revolution unfolding right before our eyes that not everyone is prepared for it. Remember the days when chalkboards and textbooks were the epitome of education?

Those days are fading and in their place, a new era is dawning, powered by Artificial Intelligence and Machine Learning. With this monumental shift, a pressing problem is slowly arising and many are left wondering, are the educational systems of today ready to harness the full potential of AI? Are our teachers equipped, our students prepared and our institutions adaptable?

This isn't just about integrating fancy tech into classrooms; it's about future-proofing education and ensuring that the next generation isn't just learning but thriving in a world interwoven with AI.

The challenges are real, from outdated curricula to a lack of understanding of what AI can offer. That is precisely why I wrote this book; to bridge the gap, highlight the problems and, most importantly, showcase the immense potential we have when AI and education join hands. This is more than just a guide, it is a call to action for all those who believe in the transformative power of education.

Imagine being a teacher, passionate and dedicated, walking into a classroom filled with eager students. You've spent hours, if not days, preparing a lesson, ensuring it's engaging and informative but as you begin, you notice something. While some students are engrossed, others seem lost, their eyes glazing over.

No matter how hard you try, reaching every student, understanding their unique needs and tailoring your approach becomes an uphill battle. The problem is the one-size-fits-all approach we are using in a world where every student is different.

Imagine yourself as a student. You're bright, curious and hungry to learn but the lessons sometimes feel disconnected from the world you live it and the technological advancements you see around you. You crave interactive, personalized learning experiences but often find yourself stuck in a traditional setup that doesn't resonate with your pace or interests.

Parents, too, feel the strain. They see their children grappling with an educational system that, in many ways, has not evolved to match the rapid pace of technological change. They worry about their children being prepared for a future intertwined with technology, especially AI.

The pain is real. The disconnect between what education could be and what it often is becomes a daily challenge. The promise of AI in education remains largely untapped and the potential benefits seem just out of reach. The question lingers: How can we bridge this gap? How can we ensure every student gets the education they need and deserve?

The adage, "If it ain't broke, don't fix it," has long been a mantra in many sectors, including education. For decades, the traditional teaching methods, the chalk and talk, the standardized tests and the rigid curricula have been the norm. Why? Because they worked—or at least, we believed they did.

But the world outside those classroom walls has been evolving at a breakneck pace and the old ways are quickly becoming obsolete. A monumental shift is on the horizon, reshaping the very fabric of education.

For the longest time, the education system was designed to cater to the industrial age. It was about producing workers with specific skills, ready to fit into well-defined roles but we are no longer in the industrial age my friends. We are in the age of information, technology and AI. The demands of today's world are vastly different and sadly, our educational frameworks have been slow to catch up.

Now, for the BAD NEWS: If we continue down this path, clinging to outdated methods and resisting change, we risk creating a generation ill-equipped for the future. An age might find them outpaced and outmatched by the very technology that was supposed to aid them.

So, who stands to WIN and LOSE in this impending shift? The winners will be those who adapt, recognize AI's potential in education and are willing to evolve. They will be the institutions, educators and students who adopt the change and harness the power of AI to redefine learning.

On the flip side, the losers will be the ones who resist change, cling to the old ways and who, despite their best intentions, find themselves left behind in an ever-evolving world.

• • •

The stakes are high, and our choices will shape the future of education for generations to come.

• • •

We've discussed some heavy stuff, but here comes the silver lining—the GOOD NEWS! With AI at its helm, the future of education is not just promising, it is transformative.

Picture a world where every student, regardless of their pace or style of learning, gets a personalized education. A world where administrative tasks do not bog down teachers so they are free to do what they do best—inspire and educate! Imagine constantly evolving institutions that are not just centers of learning but hubs of innovation.

Now, I'm not just spinning a fairy tale here. This is the potential reality we are inching towards each and every day. The best part is that you can be a part of this transformation! You can be among the WINNERS, who recognize the winds of change and adjust their sails accordingly.

Imagine if you are a teacher and instead of spending hours grading papers, an AI system does it for you, giving you more time to interact with your students and address their individual needs. Visualize yourself as a student and instead of a doing everything your peers are doing, you have a learning path tailored just for you—one that keeps you engaged and curious. As a parent, you get insights, real-time updates on your child's progress and tools to support their learning journey—all at the click of a button or swipe of a screen.

These are not pipe dreams. They are entirely possible. The goal we are working towards is a holistic, AI-driven education system that caters to the needs of this century. To get there, we need to adopt "The Big Shift". This shift means moving away from rigid, traditional methods and into flexible, tech-driven approaches. It is recognizing that the old playbook won't work anymore. We are on the cusp of a revolution and our choices will determine whether we thrive in this new era or get left behind.

The question is, are you ready to be a part of this exciting journey? To not just witness but actively shape the future of education? The ball is in your court. Let's make it happen!

ABOUT THE AUTHOR

You might be wondering, "Why should I trust this author? What makes them an authority on AI and education?"

I am Dr. Lawrence Muganga, a proud Ugandan-Canadian scholar dedicated to pursuing transformative education in this digital era. My journey has been one of passion, exploration and a deep commitment to understanding the nuances of how we learn and how we can do it better.

From the streets of Uganda to the serene campuses of Canada, my experiences have shaped my perspective on education. I've always believed that traditional teaching and learning methods are ripe for disruption, especially in a rapidly changing world reshaped by AI and Machine Learning technologies.

As a digital economy educationist, I've explored and experienced how the digital landscape alters the fabric of our educational systems. This is not just about integrating technology into classrooms; it's about re-imagining how to make learning more relevant, engaging and future-ready.

My role as an academic administrator has given me a unique vantage point. I've been at the helm of decision-making, witnessing the challenges institutions' have adapting to the digital age. More importantly, I have also seen the immense potential and promise that lies ahead if we embrace change.

Writing has been my way of sharing my insights, learnings and visions with the world. As an author, I've endeavored to put forth my ideas, hoping to spark conversations, inspire educators and drive change. My writings reflect my deep-seated belief in the transformative power of AI and Machine Learning. These are not just buzzwords, they are tools that, if wielded right, can revolutionize how we educate our future generations.

My academic journey began at the esteemed Makerere University in Uganda, where I immersed myself in Social Sciences. There, I earned my Bachelor's degree in Economics, delving deep into the intricacies of economic theories, models and their real-world applications. My thirst for knowledge didn't stop there however. I pursued a Master of Arts in Economic Policy Management from the same institution, further honing my skills and my understanding of economic strategies and their impact on policy-making.

My academic pursuits then led me to the University of Alberta in Canada. It was a period of intense learning and self-discovery. I started a Master's program in education, specializing in Adult Education. This program gave me insights into the unique challenges and opportunities of educating adults, emphasizing the importance of lifelong learning.

After dedicating years to research and study, I received my Ph.D. in Educational Administration and Leadership from the esteemed University of Alberta. The knowledge and insights I gained during this period laid a strong foundation for my subsequent endeavors. Building on this foundation, I pursued a post-doctorate program at the esteemed Derek Bok Center for Teaching and Learning at Harvard University. Honing my pedagogical skills, I culminated with a Higher Education Teaching Certificate which further equipped me with the skills and understanding to navigate the complex world of academia and contribute meaningfully to higher education.

This period of growth was pivotal, not only equipping me with a robust set of teaching tools and techniques but also deepening my ability to connect with and educate diverse students.

Beyond my administrative and scholarly contributions, I have also been deeply involved in the academic sphere as an educator. My teaching career has spanned continents, allowing me to serve as a faculty member at Victoria University and impart knowledge at various esteemed institutions in Rwanda, Canada and the United States. These teaching roles have enriched my understanding of global educational practices and affirmed my commitment to fostering an inclusive and innovative learning environment. My classrooms have been dynamic spaces where students from diverse backgrounds engage with cutting-edge content, preparing them to become agile thinkers and problem-solvers in an interconnected world.

My professional path has broadened my global perspectives, while providing me both challenging and enriching roles. In Canada, I had the privilege of serving as a Senior Policy Advisor at the Ministry of Human Services. This role allowed me to delve deep into the intricacies of governance and policy-making, shaping my understanding of how systemic changes can impact communities.

However, one of my most transformative roles was when I took the helm as the Vice Chancellor of Victoria University in Kampala, Uganda. When I first entered this position, the university was on the brink of collapse. With determination, vision and the support of a dedicated team, we managed to revive it and transform it into a beacon of academic excellence. Our student enrollment soared from a mere 243 to over 6,000 in just three short years.

But numbers only tell part of the story. I envisioned a university that was not just growing in size but also in its approach to education. With this in mind, I championed the digital transformation of Victoria University. We embraced cutting-edge technologies, revamped our teaching methodologies and became the first East African University to fully integrate digital platforms into our curriculum. This pioneering shift set us apart and positioned us as leaders in the region's educational landscape.

We moved away from mere textbook knowledge, focusing instead on equipping our students with skills and knowledge that had tangible relevance in the world outside the classroom. It has been a challenging yet rewarding endeavor and I'm proud to have shaped the university's progressive approach.

Throughout my career, I've been driven by a singular vision: revolutionizing how we approach education in the digital age.

I've ardently advocated for the digitization of higher education institutions, believing that technology holds the key to unlocking unprecedented avenues of learning. My tenure at Victoria University stands as a testament to this belief. We didn't just increase our student numbers, we transformed how we teach and learn. Traditional teaching methods, which often limited the scope and reach of education, were replaced with dynamic digital platforms. This shift wasn't just about modernization, it was about optimizing the learning experience for our students.

One of the groundbreaking changes we introduced was the adoption of online exams and digital assessments. Gone were the days of paper booklets and handwritten exams. Instead, we adopted a more efficient system in tune with the global digital trend.

My vision is much bigger than the confines of Victoria University however. I took on the mantle of a pioneer—advocating for a fully digital teaching, learning and assessment system across all universities in East Africa. It's a mission I hold close to my heart and every step we take towards this goal is a step towards a brighter, more inclusive educational future.

Beyond the academic halls and professional corridors, I've always been an avid traveler. My journey has taken me to fifty one countries, each with its unique tapestry of culture, history and experiences.

These travels have broadened my horizons and enriched my perspective, allowing me to approach my work with a global mindset. From Canada's snow-capped landscapes to Ethiopia's bustling streets, each country I've worked in has enriched my perspective and added a unique dimension to my professional narrative.

In Rwanda, I delved into the intricacies of revenue management, while in Finland, I explored the nuances of human capital development. My roles in Singapore, Sweden, the US and the Solomon Islands further expanded my horizons, allowing me to engage with various educational and developmental challenges.

At the heart of all these roles was a singular focus: human capital development. I've always believed that the true wealth of a nation lies in its people. Investing in their education, skills and well-being is the surest way to drive progress and prosperity. Whether strategizing policies in Canada or implementing educational reforms in Uganda, my mission has always been to uplift and empower individuals, equipping them with the tools they need to thrive in an ever-evolving global landscape.

I've been fortunate to receive recognition that has humbled and inspired me. One honor that holds a special place in my heart is the Pan African Revolutionary Education Award. Receiving this esteemed accolade was not just a validation of my efforts but a testament to the importance of authentic learning deeply rooted in our rich indigenous African knowledge. This award symbolized a collective acknowledgment of the need to preserve, celebrate and integrate our ancestral wisdom into modern education.

My passion for redefining education led me to write *"You Can't Make Fish Climb Trees."* Writing this book was a labor of love, culminating in years of research, observations and personal experiences. Seeing it resonate globally, garnering attention and sparking conversations, has been incredibly gratifying. The book's recognition in the educational sector reaffirms my belief that there's a pressing need to rethink and reshape how we approach learning.

Another milestone in my career was securing a grant of one hundred and three million dollars from the Bill and Melinda Gates Foundation. This wasn't just a financial boost but a vote of confidence in my vision and endeavors. With the support of such a renowned foundation, I felt further empowered to drive change, innovate and make a lasting impact in the world of education.

Looking back, I'm grateful for the myriad of opportunities I've had, the challenges I've overcome and the countless individuals I have had the privilege to work with and learn from. Each chapter of my career has been a lesson, a story and a stepping stone to the next big adventure.

In the ever-evolving landscape of education, I felt a pressing need to address the age-old, conventional education model that often binds students to the confines of physical classrooms. This led me to pen down my thoughts, experiences and vision in my book, *"Authentic University."*

"Authentic University" is not just another book on education, it is a call for transformation. It challenges the traditional norms and presents a vision for a modern educational system that is in sync with our technology-driven world. The book emphasizes real-world, practical learning, urging institutions to prepare students for the challenges and opportunities of today and tomorrow.

One of the core tenets of my book is the idea of affordable education that recognizes and prioritizes the uniqueness of every student. It advocates for personalized learning experiences, ensuring that education remains relevant and meaningful. The book underscores the importance of equipping students with practical and employable skills, which add value to their educational journey and pave the way for a smoother transition into the workforce.

While universities have remained unchanged over the decades, "Authentic University" offers a fresh perspective, highlighting the need for a paradigm shift in higher education. It's a guide, a road map and a vision for the future of education.

The feedback and reviews for "Authentic University" have been overwhelmingly positive. Phil Jackson, an esteemed reviewer, described it as an "education game-changer" that every student, parent, educator and administrator should have in their library. He praised the book for its forward-thinking approach and emphasis on the unique learning needs of students in a modern environment.

For those interested in further exploring my vision for the future of education, "Authentic University" is available globally, both in print and digital formats. You can find it at renowned book outlets like Barnes and Noble, Chapters Indigo, Amazon, Powell's Books, Partridge Publishing, Google Books and Apple Books.

The age-old system churning out job seekers needs a revamp and this book lays a blueprint for that change. At the heart of my vision is authentic learning—fostering an environment where students are not just passive recipients of knowledge but active participants in their educational journey.

Coupled with this is promoting entrepreneurial aspects, encouraging students to think outside the box, innovate within existing structures and drive change from within. It's a call to action, urging institutions to rise to the challenges of our times and sculpt a brighter education future.

My vision for authentic education is rooted in a simple yet profound belief that learning should be real. It should be tied to our world, our challenges and the opportunities that lie ahead. Theoretical knowledge, while valuable, is only one piece of the puzzle. What truly matters is how we apply this knowledge, adapt it to solve real-world problems and innovate to create a better future.

In my view, a student-centered approach is the key to unlocking the true potential of every learner. By tailoring the learning experience to each student's unique strengths, interests and aspirations, we can create an environment where learning is not just effective but deeply meaningful.

Personalized learning experiences are not just a pedagogical choice but a necessity in today's world. The demands of modern employers are evolving at an unprecedented pace. The skills and competencies they seek are not static; they change with time. To prepare our students for this dynamic landscape, we must ensure that their learning experiences resonate with the realities of the modern workplace.

I've witnessed the dawn of a new era, where Artificial Intelligence (AI) and Machine Learning (ML) are not just far-fetched concepts but powerful tools reshaping the very fabric of learning. As I navigated this landscape, it became increasingly clear that these technologies hold transformative potential, a potential that we, educators and visionaries, must harness.

My advocacy for AI and ML in education is rooted in a deep-seated belief that these technologies can bridge the gap between traditional pedagogies and the dynamic needs of the 21st-century learner. They offer us the ability to personalize learning experiences, adapt the content in real-time based on student performance and provide previously unimaginable insights.

In this book, I share how these technologies can revolutionize education and create learning experiences that resonate with the real world and prepare our students to tackle the complex challenges of the future. Readers should treat this book as both a guide and a conversation starter. In a rapidly evolving world where AI is becoming more integrated into our daily lives, especially in education, this book provides insights into AI's challenges, benefits and transformative power. It's not just about understanding the technology, it is about recognizing its impact on the human aspect of teaching and learning. It's about the synergy between humanity and technology. While AI is an incredible tool, the heart of education will always be rooted in human interactions, creativity and empathy.

This book addresses the pressing need for educators, policymakers, students and parents to navigate the promises and pitfalls of AI in education. Its balanced approach sets this book apart in a landscape saturated with information. It doesn't just sing praises of AI; it delves deep into ethical considerations, the ever-evolving role of educators and the essence of lifelong learning in an AI-driven world.

This book is a culmination of my experiences and a testament to my deep connection with the evolving landscape of technology in education. Every page is fueled by my passion for the subject, a love from years of hands-on experience, and a vision for a future where technology and education seamlessly merge to create unparalleled learning experiences.

To anyone invested in the future of education, I invite you to join me in this quest for innovation, to challenge the status quo and to envision a world where learning is not just a process but a lifelong journey of discovery, growth and transformation.

FOREWORD

Etched in our collective memory, 2020 will forever be remembered as the year in which our world, through loss and adversity, changed beyond recognition with the advent of the COVID-19 pandemic.

When the World Health Organization (WHO) declared the end of COVID-19 as a global health emergency in May 2023, the world did not seem to take much notice. After all, by then, daily life had generally resumed with some semblance of normalcy. One also suspects that so much change had happened in such a short span of time—the ongoing war in Ukraine and the ensuing crisis of inflation, sovereign debt crises and the shift in a long-established geopolitical status quo, to only name a few—that the WHO's declaration was easily overshadowed.

One such disruptive change was OpenAI's ChatGPT, an Artificial Intelligence (AI) language model capable of generating human-like text based upon direct prompts from users. Whilst it had been launched as early as November 2022, it was still a recurring theme in the news and daily conversations by the time COVID-19 was declared to be officially over. Notwithstanding that the development of AI had been discussed for decades and that machine learning had already been in use—albeit behind-the-scenes—for years, this was the first time that most of us were able to directly witness, engage with, and understand the implications. Without a doubt, the age of AI was now upon us—for better or for worse.

It was only a matter of time before we found ourselves bound to discuss the consequences that this new technological development would have on every part of our societies—from the risks of its potential use for military purposes to the adverse consequences it could have on employment prospects, passing by the possibility of enhancing the general quality of life by leaving tedious tasks to software, and the prospective life-saving uses in the health sector through pro-active and constant monitoring of vital signs.

Needless to say, the education sector was not spared from this pervasive conversation topic and soon enough—to simplify—two opposing camps appeared. The dystopian camp perceived such advancements as an existential, dehumanizing threat for education systems, while the utiopian camp consisted of those who saw an opportunity to exponentially enhance educational quality that needed to be accepted and adapted to the fait accompli. The dystopian camp saw a mindless and dangerous herd mentality. For the former, the education system needed to reflect society and its changes, and for the latter, it needed to remain impermeable to a passing fad and preserve its role as a forger of society. Two, seemingly irreconcilable premises—and, when faced by opposing positions, truth has the most irritating habit of refusing to reveal itself, leaving partisans to battle it out until one camp emerges as the victor. Even then, such victories are left to time to decide, with the benefit of retrospect, on whose side truth had fought—or, indeed, whether it even had taken a side at all.

In any event, for us mere mortals, neutrality would be irresponsible. For we can afford neither infertile indifference nor hesitant inaction: the debate must be had, as thoroughly and participatory as possible. The stakes are high, and our duty is all the greater, when we speak of education: it is the future of society, not merely of the classroom, that is on the line.

Should I be asked for my own views on the matter, neither dystopian nor utopian, I would invariably lean towards a third way.

Education systems do not exist in a vacuum. There is no such thing as education in opposition to the 'real world', which is in a constant process of change. They are inextricably linked, mutually influencing each other. One may decide to disregard such societal changes, continuing with business as usual, but that does not alter the inevitable: these changes occurring outside the formal education system will impact it, one way or another. The only issue, then, is for us to prepare for the manner in which such developments will affect the system. If we refuse to do so, then we will have lost the opportunity to harness their positive effects, and we will have amplified their adverse consequences.

Moreover, learning is not—and never has been—confined to the four walls of the classroom. From birth, children—and, for that matter, everyone—constantly learn with everyday social interactions, through every text or video encountered, in every life event experienced. When our education systems, as they have historically operated, deny and reject this reality, they not only announce their self-induced obsolescence, but also commit the injustice of impoverishing the formal learning process. The only realistic and—I dare say—desirable way forward is for our systems to conjugate the formal educational program in the classroom with the informal learning process occurring outside of it, in society. To paraphrase and build upon Paulo Freire's seminal 'Pedagogy of the Oppressed', our educational institutions must become a space of communion and dialogue in and with the world.

In the specific case of AI, much ink has been spilled—or, more appropriately, much digital data has been generated—on its dangers, with concerns coalescing on the perceived spread of cheating. This is a legitimate concern which, in fact, could constitute a crisis for the system if materialized. The good news is that crises are also opportunities. In this instance, it is the opportunity to finally question our long-held, inherited assumptions about learning and transform our education systems in consequence. Regarding cheating, rather than limiting ourselves to the debate on its avoidance, we should instead ceaselessly poke at the institutional structures and processes that Incentivize students to cheat—or, in other words, a system in which students are made to be more interested in passing than in learning.

Worse still, if the measure of the critical thinking skills that we seek to nurture in our youth relies on whether they are able to find the right answer to the problem we present to them—which, perhaps, AI could become more capable of doing than any human being—then we must urgently alter our course. The true measure of intelligence should not only be knowing the right information, but also knowing how to accurately problematize the situation; it should not merely be in finding the right answer, but equally in articulating the right question. To blame AI for increased cheating is too facile an exercise—we must accept our own responsibility in the matter: the debate should rather center on how to ensure that our education systems are designed to incentivize learning over passing, and move from problem-solving to problem-posing approaches.

Such an entreaty to the dystopian camp notwithstanding, one cannot simply jump on the bandwagon of technology—and, in this particular case, of AI —for its own sake. It is true that technology is a tool, and tools are neither good nor bad in and of themselves: the knife should neither be condemned for a stabbing, nor praised for a meal. Yet concluding that, because we have been living through the 4th Industrial Revolution, we must now just embrace technology, in all its forms and in all sectors, seems profoundly misguided.

It was not so long ago that, in numerous influential circles of self-proclaimed thought-leaders, we were continuously told that the future of education would be borderless and hence (the thinking went) more inclusive. Putting aside, for the sake of argument, the persisting techno-digital divide, the COVID-19 pandemic demonstrated the limits of what was once touted, in utopian terms, as a desirable future: whilst virtuality did enable the continuity of schooling (for those who had connectivity) at a time when classrooms had to be closed, it also impoverished learning. I personally recall that as a student, whilst listening to a teacher—from whom I learned—the classmate sitting next me muttered a comment to himself, and I, upon overhearing him, had a thought that I would otherwise not have had on my own—and I learned that way as well. Learning is clearly as much of a social process as it is a cognitive one. This social dimension of learning is entirely lost through virtuality. When distance learning enables greater access and exposure, then it is undeniably necessary; when, however, it is adopted for its own sake, it is equally undesirable. Hence, whilst it may be that it is no longer a question of "whether" to use technology, we must still ask ourselves "why?", "when?", and "how?" to use it in the education system.

In addition, we must also have the lucidity to realize that not all tools are created equal. Whereas a knife merely changes the way we perform a task, technology has the potential to shape the way that we understand and how we are in, the world. Despite appearances, technology is neither socially nor culturally neutral: its development is a human-led process, during which the programmers—who naturally have biases, socio-cultural interpretative frameworks and specific, individual goals—inevitably transfer non-neutral properties to the digital tool which, in turn, shapes us, the users.

In the case of AI, this problem is further exacerbated, not only because the data on which it is trained will probably contain undetected biases, but also since even well-intentioned attempts to limit such biases during its design can lead, given its generative nature, to unintended consequences. As I write this, the news on technology has been dominated by Google DeepMind's Gemini, with the AI image generator producing historically inaccurate images of ethnically-diverse US Founding Fathers and Nazis—and the company has had to acknowledge that it would need time to resolve the issue, given the complexity of the model. In other words, not only does AI contain biases that may go unnoticed and produce false information qualified by the industry as "hallucinations", but it also appears that the complexity of these neural networks makes any swift—let alone comprehensive—correction challenging. To wholeheartedly accept the introduction of AI within our classrooms, merely because it exists, is too stoic an approach to make for good policy. In addition to asking "why?", "when?", and "how?" to use technology and AI in education, our education systems should also ask "which ones?" and "designed by whom?"

It is hence not unreasonable to hold the view that both the dystopian and the utopian camps have benefited from truth's indecision. Either side may win today's battle, but it is with time and the benefit of retrospect that we shall probably come to learn that truth had not chosen a side that day, and that it was as much a victor as it was a casualty. If we are to purposefully avoid this outcome, a third way must emerge—the conciliatory camp: not whether to use technology and AI in our education systems, but under what conditions and frameworks to do so; not limiting ourselves to the dichotomy of having education either simply ignore or docilely comply with society's changes, but conjugating the present with the future.

In authoring this book, Dr. Lawrence Muganga has clearly chosen this very path: whilst his optimism shines at every line, he has not declared himself for the jubilant, utopian camp; whereas his understanding of the risks emerges throughout the pages, his is not of the resigned, dystopian camp either. Yet my foreword does not do justice to the following chapters, for the author has been bolder and more daring: with the commitment to education that many have known to be his for years, he has gone beyond asking questions and sought to lay out a concrete, actionable way forward—not only to mitigate an inevitable crisis, but also to make an opportunity out of it for the long-overdue transformation of our education systems.

Naturally, readers may not agree with all of his assessments, nor concur with every single one of his proposals. They may, in fact, vehemently disagree more than once. For in the case of education, the most human of sectors, there can be no recipes nor absolute certainties, only continuous endeavors for betterment that must obey the essential precept of contextualization. The value of this book is not to be found in its being a roadmap to be applied thoughtlessly. On the contrary, its strength lies in articulating tangible proposals as food-for-thought, in reconciling the utopian and the dystopian camps, and in doing so in a digestible structure and accessible language which will enable educators and lay citizens to partake, alongside policy-makers and educational institutions, in this crucial debate—one that has too long been reserved for the exclusive, jargon-riddled circles of experts—on the future of education and, by extension, of our world.

Régis Debray's "Mediology" accurately points us towards the unsavory realization that technical revolutions are more durable than political or, indeed, social ones: after an Allende, one may suffer a Pinochet; after the right to universal and free healthcare, one may be subject to exclusionary privatization; but after the computer, one simply does not revert to the typewriter.

The 4th Industrial Revolution has been with us for years, and we have now entered the age of Artificial Intelligence. Pandora has opened the box—there is no closing it back. What is left for us is the resolve to come together, debate the matter thoroughly, and determine decisively the rules of engagement, policy frameworks and regulatory guardrails that will, one hopes, enable us to use this developing technology for the sake of building the education that we need and of shaping the future that we want.

In this regard, Dr. Muganga's book constitutes a welcome entry-point to this essential topic and a most valuable contribution to the ongoing debate.

**His Excellency Sheikh
Manssour Bin Mussallam**

Organisation of Southern Cooperation,
Secretary-General, 2021-2027
Addis Ababa, Ethiopia
3rd March 2024
at 21:30

PART 1

FOUNDATIONS AND APPLICATIONS OF AI IN EDUCATION

1
THE FOUNDATIONS OF AI IN EDUCATION

HISTORICAL FOUNDATIONS

When we delve into the origins of AI, Alan Turing invariably stands out as a pivotal figure. Turing's work in the early 20th century, when computers were mere figments of science fiction, laid the cornerstone for AI. His concept of a universal machine in the 1930s, capable of simulating any mathematical computation representable as an algorithm, planted the seed for today's computational landscape.

Turing's most significant contribution perhaps came in the 1950s with his probing question, "Can machines think?" This inquiry led to the Turing Test, a benchmark for evaluating a machine's ability to exhibit human-like intelligence. More than just a theoretical exercise, Turing's work ignited a paradigm shift.

Now, you might wonder about the relevance of these historical insights to education. The answer lies in the appreciation of AI's evolution. Understanding the genesis of AI enriches our perspective on its potential applications in various fields, especially education. This historical context helps us recognize AI's transformative capacity in educational settings, from traditional classroom approaches to innovative, tech-driven learning experiences.

THE FIRST WAVE OF AI

Think of the first wave of AI not as a scene from a sci-fi movie with sentient robots but more like the beginning of a journey with "Symbolic AI" or "Good Old-Fashioned Artificial Intelligence" (GOFAI). This phase was about teaching machines to interpret and process information using symbols, akin to giving a child building blocks and a manual to construct a tower. The era saw significant developments, from problem-solving to chess-playing machines, sparking curiosity and wonder about the potential of machines.

In education, this era hinted at a future where machines could guide students through problems, much like a digital tutor. The dawn of Symbolic AI in education was about envisioning a more interactive and personalized learning experience.

AI IN THE LATE 20TH CENTURY

NEURAL NETWORKS AND EXPERT SYSTEMS

Moving past the initial wave, the AI community shifted to "neural networks," an endeavor to design machines that imitate the human brain. Unlike Symbolic AI, neural networks learn from data, emulating how we gain knowledge through experience.

Simultaneously, "expert systems" emerged, aiming to replicate the decision-making abilities of human experts. These systems, precursors to today's recommendation algorithms, brought the possibility of tailored educational guidance similar to having a digital expert by a student's side.

However, this burst of innovation led to overly optimistic predictions about AI, culminating in the "AI winter," marked by reduced interest and funding. This phase served as a reminder of the complex journey of AI development.

THE RESURGENCE OF AI

BIG DATA AND ADVANCED COMPUTING

The skepticism following the AI winter was soon replaced by optimism as we entered the era of big data. This wasn't just about extensive datasets but the dawn of a new age in data-driven insights. From online behaviors to social media patterns, data has become a crucial resource.

In an educational context, big data enabled a profound understanding of each student's learning journey. When paired with advanced computing, the wealth of data allowed AI to offer tailored educational experiences similar to transitioning from a basic guidebook to a dynamic, interactive learning system. This evolution marked a significant stride in making education more personalized, engaging and effective.

The narrative doesn't stop there however. We must shift focus to the African classroom, envisioning how these technological advancements could transform educational experiences where less advanced methods have been the norm. Doing so, we consider the unique challenges and opportunities within this setting, emphasizing the potential for AI to offer solutions that resonate with local educational needs.

Through this lens, the resurgence of AI becomes a story not just about technology but about unlocking the potential of every student. It's about bridging the gap between traditional educational methods and today's dynamic, data-rich world.

UNDERSTANDING AI AND MACHINE LEARNING

BASICS OF AI: NARROW AI VS. GENERAL AI

Narrow AI excels in specific tasks like the technology in your smartphone's voice assistant or your favorite streaming service's recommendation algorithm. It's specialized, like a chef who masters one type of cuisine. On the other hand, General AI, which remains an aspiration, would be akin to a renaissance individual adept in multiple fields—from culinary arts to poetry.

MACHINE LEARNING VS. PROGRAMMING

THE PARADIGM SHIFT

Traditional programming involved feeding computers rigid, step-by-step instructions, where earning was done through a fixed set of rules and processes. Machine Learning, on the other hand, revolutionized this by allowing computers to learn and improve from data, similar to a person mastering a skill through practice and experience.

DEEP LEARNING

THE MAGIC OF NEURAL NETWORKS

Deep Learning, enabled by neural networks, mirrors the complexity of the human brain. It powers the technology behind face recognition in social media and voice recognition in digital assistants. This technology is learning from vast amounts of data to make intelligent decisions, resembling how we learn and interpret the world.

SUPERVISED VS. UNSUPERVISED LEARNING

CLASSROOM ANALOGIES

In supervised learning, imagine a classroom where the teacher guides every step, similar to how we provide data with known outcomes to computers. This approach helps the machine predict and learn, comparable to a student becoming independent after guided learning.

Unsupervised learning, in contrast, is like exploratory learning without explicit directions, where students independently discover patterns and ideas. This mirrors how computers in unsupervised learning find hidden structures in data without predefined answers, fostering an environment of discovery and innovation.

HOW MACHINE LEARNING ALGORITHMS WORK

Teaching a machine learning algorithm is like guiding a child to distinguish between different objects. With examples (data), the machine, much like a child, learns to differentiate and recognize patterns. The process involves trial and error, refining its understanding over time.

Consider machine learning as a culinary art. The data serves as ingredients, the algorithm as the recipe and the computer as the chef. The chef combines ingredients following the recipe, refining the dish through tasting and adjusting, analogous to the computer optimizing its learning process. Supervised learning is like labeling each ingredient, while unsupervised learning is akin to a chef identifying and utilizing ingredients without labels.

AI TOOLS AND TECHNOLOGIES

NATURAL LANGUAGE PROCESSING (NLP)

NLP enables our devices to understand and respond to human language. This technology powers chatbots for customer service and language learning apps. NLP has evolved search engines, making them contextually aware and understanding nuance and colloquialisms.

In education, NLP transforms language learning, moving from rote memorization to interactive, AI-assisted methods. Students can now practice pronunciation and grammar interactively, as if they have their very own personal tutor sitting by their side.

COMPUTER VISION

AUTOMATED GRADING AND INTERACTIVE LEARNING

Computer vision allows computers to interpret visual information, similar to human vision. It is used in smartphones for facial recognition and in-app identification. In education, this technology can automate grading by reading and interpreting students' handwriting, accommodating various styles and nuances.

Furthermore, computer vision can enrich interactive learning. Biology classes, for instance, could use 3D models for a more engaging educational experience, with real-time feedback and dynamic adjustments. This technology opens up possibilities for dynamic, adaptable learning environments catering to diverse learning styles.

PREDICTIVE ANALYTICS

ANTICIPATING STUDENT NEEDS AND OUTCOMES

Predictive analytics uses historical data, statistical algorithms and machine learning to forecast future outcomes. This tool helps identify students needing extra support by analyzing various data such as attendance, grades and socio-economic factors, enabling early intervention. This technique can also customize teaching methods based on historical success patterns, like integrating visual aids and interactive activities shown to enhance learning.

Drawing a parallel to familiar technologies, predictive analytics in education is likened to how streaming services recommend content based on viewing history and preferences. It evaluates a broad spectrum of variables to suggest suitable educational resources or interventions for students, similar to how these platforms tailor suggestions to individual viewers.

Predictive analytics extends beyond student needs, forecasting enrollment trends and program success, aiding institutions in resource allocation and long-term planning.

ROBOTICS IN EDUCATION

FROM PROGRAMMABLE TOYS TO HUMANOID TEACHER ASSISTANTS

Robotics in education has evolved from simple programmable toys to sophisticated humanoid teacher assistants. These advancements transform traditional learning environments by introducing dynamic, interactive elements.

- **Programmable Toys for Learning:** These advanced robotic tools teach students coding and computational thinking in a hands-on manner. For instance, coding a robot to navigate a maze can bring abstract programming concepts to life, enhancing problem-solving skills.
- **Humanoid Teacher Assistants:** These robots, emerging in some classrooms, support teachers and offer personalized assistance to students. They adapt to individual learning styles, providing a unique and engaging educational experience.
- **Interactive Learning with Robotics:** Robotics in education goes beyond conventional teaching methods by offering immersive, interactive learning experiences. This is particularly impactful in STEM education, where hands-on learning significantly enhances comprehension and retention.
- **Fostering Future-Ready Skills:** Early exposure to robotics equips students with essential skills like problem-solving, creativity and critical thinking. It also nurtures soft skills such as teamwork, adaptability and effective communication, preparing students for a technologically advanced future.

Incorporating robotics in educational settings is a strategic approach to enriching learning experiences and equipping students with skills for a tech-centric world.

KEY TAKEAWAYS

Historical Significance: Understanding the historical context of AI, from Turing's foundational work to the first wave of symbolic AI, provides insight into AI's current capabilities and future potential in education.

Evolution of AI: The shift from symbolic AI to neural networks and expert systems marks a significant advancement in creating more adaptive and personalized learning experiences.

Modern AI Applications: The resurgence of AI, fueled by big data and advanced computing, has led to innovative educational tools, such as predictive analytics and robotics, transforming the learning landscape.

CHAPTER LEARNING ACTIVITIES

ACTIVITY 1: AI TIMELINE PROJECT

- **Objective:** Create a visual timeline of the key milestones in AI development mentioned in the chapter.
- **Activity:** Research additional sources to expand on the milestones provided in the chapter. Then, design and illustrate a timeline that includes these milestones and their significance to the field of education.

ACTIVITY 2: PREDICTIVE ANALYTICS EXPERIMENT

- **Objective:** Understand how predictive analytics can be used to enhance learning outcomes.
- **Activity:** Collect data from a mock classroom scenario (this can be based on past educational experiences or created data). Use simple statistical tools to identify patterns and make predictions about student performance.

ACTIVITY 3: ROBOTICS IN EDUCATION ROLE-PLAY

- **Objective:** Explore the potential impact of robotics on education through a role-play exercise.
- **Activity:** Divide into groups, each representing different stakeholders in the educational process (students, teachers, administrators). Role-play a scenario where a humanoid assistant is introduced into the classroom, discussing potential benefits and challenges.

2
PERSONALIZED LEARNING WITH AI

PERSONALIZED LEARNING DEFINED

BENEFITS FOR LEARNERS

Picture a classroom where every student has a unique learning path tailored to their strengths, weaknesses, interests and pace. Sounds dreamy, right? That's personalized learning for you. It's all about making learning a custom-fit experience rather than a one-size-fits-all approach.

Think of personalized learning as a bespoke suit. Instead of buying off-the-rack, you get something that fits you perfectly, accentuating your best features. That's what customized learning does for students; it accentuates their strengths and supports their weaknesses.

So, what are the benefits? For starters, students are more engaged because they are learning in a way that resonates with them. They feel valued, understood and, most importantly, empowered. It's like giving them the keys to their educational journey. Plus, with personalized feedback and resources, they can tackle challenges head-on and celebrate their successes. It's a win-win!

THE CONCEPT OF PERSONALIZATION

Personalization in education is a powerful approach that recognizes and celebrates the uniqueness of each student. At its core, it acknowledges that every student is an individual with a distinct background, culture, experiences, learning styles and goals.

Personalization begins with acknowledging that no two students are exactly alike. Some students may excel in certain subjects while needing extra support in others. Personalization takes all of these differences into account and tailors educational experiences accordingly.

Think about your favorite coffee shop where the barista knows your order by heart. This small but thoughtful gesture makes you feel special and valued. In education, personalization is akin to that barista knowing your unique preferences. It's not just about delivering a standard menu, but about creating an educational experience that resonates with each student's needs and preferences.

Personalization goes beyond merely customizing content or pacing. It delves into the world of emotions and human connection. It is about making learners feel seen, heard and understood. When students feel their individuality is acknowledged, it can profoundly impact their motivation, engagement and overall learning experience.

Personalization also involves adapting teaching methods to cater to different learning styles. Some students may thrive in a visual learning environment, while others prefer hands-on activities. Personalized education recognizes these differences and offers a variety of approaches to ensure that all students can learn in ways that resonate with them.

Personalized learning isn't a new concept. It's been around for centuries. Remember the one-room schoolhouses where teachers catered to students of different ages and abilities? Or the ancient Greek tutors who provided one-on-one instruction? That was personalized learning in its early form.

But while the concept has been around for ages, the methods have evolved. With the advent of technology and AI, we have the tools to take personalization to new heights. Think of the transition from handwritten letters to instant messaging. The essence of our interpersonal written communication remains the same, but the medium has drastically changed.

THE FOUR PILLARS

Imagine a grand, ancient temple standing tall and proud. Now, this temple is supported by four mighty pillars, each unique yet equally important. These pillars are the essence of personalized learning.

1. **Differentiation:** Think of this as the chef's special. Just as a chef tweaks a dish based on the diner's preferences, educators modify content based on students' needs. It's all about serving up lessons that resonate.

2. **Individualization:** Remember those choose-your-own-adventure books? Individualization is just that. It's letting students choose their learning pace and even some content.

3. Adaptation: This one's like a GPS for learning. If a student veers off course, adaptive tools recalibrate to get them back on track. It's dynamic, responsive and oh-so-smart!

4. Personal Learning Paths: Imagine a playlist for learning. Students get a curated list of lessons tailored just for them. It's like having a personal DJ for education!

PERSONALIZED LEARNING BENEFITS

When comparing traditional learning to personalized learning, it is essential to understand the nuances and implications of each approach.

Traditional Learning: This is the "one size fits all" education model. It is the old, comfy sweater analogy—reliable and familiar but not necessarily suited to everyone's unique shape and size. The curriculum is standardized in traditional classrooms, meaning all students receive the same instruction, regardless of their learning styles, interests or pace. The teacher is the central figure, imparting knowledge to students who are expected to absorb and replicate it. Traditional learning has its merits, such as a structured environment and a uniform metric for measuring student progress, but it can also be rigid and, as mentioned, impersonal.

Personalized Learning: On the flip side, personalized learning is the tailor-made suit—carefully measured and cut to fit the individual perfectly. This approach is designed with the learner at the forefront. It's equivalent to a curated Spotify playlist, where every song is chosen based on your tastes instead of a generic cassette tape with a fixed set of tracks.

BENEFITS OF PERSONALIZED LEARNING OVER TRADITIONAL LEARNING:

- **Engagement:** Personalized learning strategies are designed to increase student engagement by presenting materials and activities that are relevant and interesting to them. When students see the value and purpose in their learning, they are more likely to be motivated and involved.

- **Retention:** Tailoring the learning experience to the individual can lead to better retention of information. When students learn in the best ways for them, they're more likely to understand and remember the concepts taught.

- **Pacing:** Personalized learning allows students to progress at their own pace. Fast learners can move ahead to more challenging material, while those who need more time can take it without feeling rushed or left behind.

- **Technology Integration:** Personalized learning often leverages technology to create adaptive learning paths for students. Educational software can adjust the difficulty of tasks in real-time based on student performance, providing a truly individualized learning experience.

- **Deeper Understanding:** Personalized learning facilitates a deeper understanding of the material by focusing on students' interests and needs. Students are encouraged to explore subjects in ways that connect with their existing knowledge and experiences, leading to a more profound grasp of new concepts.

- **Skill Development:** Personalized learning often emphasizes the development of critical thinking, problem-solving and self-directed learning skills. These are crucial for success in the 21st-century workplace.
- **Teacher's Role:** In personalized learning, the teacher shifts from the sole source of knowledge to a guide or coach. This allows for more meaningful interactions between teachers and students, as teachers can provide support and feedback specific to each student's needs.

ADAPTIVE LEARNING PLATFORMS

Remember those days in school when you would grasp a concept instantly, but your friend would need a bit more time or vice versa? We all learn at different paces and that's perfectly okay. Sadly, traditional education systems often miss the mark in catering to these individual needs.

Enter adaptive learning platforms. Imagine a digital tutor who's always by your side, understanding your strengths, weaknesses and learning pace. As you interact with the platform, it's constantly learning more about you. Did you miss a question about fractions? No worries! The platform will provide additional resources or exercises to help you grasp the concept. Aced a topic? Great! The forum will offer advanced materials to challenge you further.

It is like having a GPS for learning. Just as a GPS reroutes you based on real-time traffic conditions, adaptive learning platforms adjust the content based on your real-time performance. If you're breezing through a topic, it will accelerate. If you're struggling, it'll slow down, offering additional support. But these platforms aren't just about adjusting the pace. They're about understanding the 'why' behind your responses. Did you get a math problem wrong because of a calculation error or a conceptual misunderstanding? Depending on the answer, the platform will tailor its feedback.

For teachers, this is a game-changer. Instead of spending hours identifying individual student needs, the platform provides insights, allowing educators to offer targeted support quickly, ensuring that no student is left behind.

In a world where we are moving towards personalization in almost every aspect of our lives, from movie recommendations to product advertising, isn't it time our education system caught up? With adaptive learning platforms, we're taking a giant leap in that direction, ensuring every student gets the personalized attention they deserve.

USING STUDENT DATA TO TAILOR LEARNING EXPERIENCES

I often find myself reminiscing about my school days, especially when I'd gaze out of the window, lost in thought, while the teacher droned on about something I could not connect with. I would think, "If only they knew what I was feeling or struggling with." Fast forward to today, and my wishful thinking wasn't so far-fetched.

Picture a world where educators have a magic crystal ball that lets them peek into the minds of their students. A world where they can instantly know who's struggling, who's bored, and who's ready for more. That is the power of data-driven insights.

Every time students interact with a digital platform, whether answering a quiz, watching a video or even spending time on a particular topic, they leave behind digital footprints. These footprints, when analyzed, paint vivid Images of the student's learning journey. It is like reading a diary, but we have data points instead of words. Now, here is where it gets interesting. By harnessing this data, educators can tailor learning experiences to fit the unique needs of each student. No more one-size-fits-all; it's all about what fits YOU.

Let's dive a bit deeper. Imagine a student named Mia. She's acing algebra but struggling with geometry. Traditional teaching methods might not pick up on this nuance, but with data-driven insights, her teacher can see the exact areas where Mia needs support. Maybe she needs more visual aids or a different approach to understanding shapes. The possibilities are endless, and they're all rooted in data.

But it's not just about identifying weaknesses. It's also about celebrating strengths. If a student is passionate about a particular topic, why not provide them with advanced materials? Let's nurture that spark and turn it into a roaring flame.

Now, I know what you're thinking. "Is it ethical to use student data?" Absolutely, but with caution. It is paramount that data is used responsibly, ensuring privacy and security. After all, great power comes with great responsibility (we will discuss more about it in the section below).

CHALLENGES AND ETHICAL CONCERNS

DATA PRIVACY

I recently chatted with my niece about her favorite toy—a smart doll that can chat, sing and tell stories. It was all fun and games until she mentioned, "My doll knows my favorite color and bedtime stories!" That got me thinking. In a world where even toys collect data, what does it mean for our students in schools and universities?

We live in a world where every click, every answer, every hesitation can be recorded. It is happening right now on platforms like Facebook. Now translate that to learning platforms. With the rise of AI, the amount of student data being collected is staggering. And while this data can be a goldmine for personalized learning, it also opens a Pandora's Box of privacy concerns.

I'm all for innovation in education but the truth is, we tread on thin ice regarding our students' data. Think about it. This isn't just about grades or attendance. We're discussing learning styles, strengths, weaknesses and emotional responses. This data could be misused in the wrong hands, leading to bias, discrimination or exploitation.

Remember the story of the boy who cried wolf? Now, imagine if every time he called, there was a record. Over time, one could predict exactly when he would cry next. Similarly, with enough student data, there's a risk of predicting and possibly misjudging a student's potential based on their past. It's like being forever labeled because of one's history.

Back to our main point, though—privacy. In the world of big data, data breaches is a pressing concern. We've all heard of instances where personal data was hacked, leading to identity theft and other nightmares. Now, imagine that happening with our students' data. The implications are chilling.

While these concerns are real and pressing, they're not insurmountable. We can navigate these choppy waters with robust data protection policies, transparent data collection practices and a commitment to ethical AI. It's about striking a balance between innovation and privacy.

ALGORITHMIC BIAS

HOW IT CAN AFFECT PERSONALIZED LEARNING OUTCOMES

So, I was binge-watching this sci-fi series the other night, where robots made decisions for humans. It was pure entertainment on my part, until one robot made a decision based on flawed data. That's when it hit me—aren't we as educators, in some ways, letting algorithms decide for our students? And what if these algorithms are biased?

Imagine you're baking a cake. You've got a recipe passed down through generations. But if you start with salted butter instead of unsalted, your cake's taste changes, right? Similarly, the data you feed into an AI algorithm determines its output. If the data is biased, the results will be too. It's like the old saying, "Garbage in, garbage out."

Now, let's relate this to education. Personalized learning, driven by AI, promises tailored educational experiences for each student. But, if the algorithms driving this personalization have biases, we might end up pigeonholing students or, worse, perpetuating existing stereotypes.

I once had a friend, Maya, a brilliant artist, who was told she'd be better off in accounting. Why? Because the test was biased toward certain professions. Now, amplify this with AI, and you see where I'm going. Back on track. The danger of algorithmic bias in education is that it can limit opportunities, narrow down potentials and even misjudge capabilities. A biased algorithm might steer a budding poet toward engineering or push a future scientist into the arts. It's like having a slightly off GPS; you might end up at the wrong destination.

Here's the silver lining though—recognizing the problem is half the battle. We can work towards more equitable algorithms by being aware of these biases. Ensuring continuous checks, balances and tweaks to keep everyone on course.

EQUITY IN PERSONALIZED LEARNING

Equity in personalized learning is a critical issue that ensures that all students have an equal opportunity to benefit from modern educational approaches. While some schools and students may have access to advanced AI-driven learning platforms, others may not have the resources to afford such luxuries. This creates a gap in the quality of education and opportunities for different groups of students.

It's crucial to distinguish between equity and equality. Equality means treating everyone equally, while equity means giving each individual what they need to succeed. Equity ensures that every student, regardless of background, has access to the resources and support necessary to thrive.

Technology is a powerful enabler for personalized learning, but it's not a silver bullet. It's not enough to provide high-tech tools and expect equitable outcomes. The human element, including teachers and the support system, is equally vital. Technology should enhance, not replace, the human touch in education.

To bridge the equity gap in personalized learning, it is essential to recognize that technology alone won't solve the problem. It requires a holistic approach that blends technology with a human touch. This means providing professional development for teachers, creating inclusive curricula and ensuring that technology benefits all students, irrespective of their socio-economic backgrounds.

Inclusivity is a key goal. This means designing educational systems and tools that all students can access and utilize, regardless of their abilities, resources or challenges. This may involve making technology more accessible and accommodating diverse learning needs.

TRANSPARENCY AND ACCOUNTABILITY

Imagine you're at a magic show, and the magician pulls off an incredible trick. You're left in awe, wondering, "How did they do that?" Imagine if AI in education was like that magician, performing wonders but leaving everyone in the dark about its methods. It's not so magical anymore, right?

Transparency is our first stop. Just like you'd want to know how that magic trick works (well, some of us would!), it is essential to understand how AI makes decisions in the classroom. If an AI tool recommends a particular learning path for a student, educators and parents should know why. No smoke and mirrors here!

Remember when you were a kid, and you'd shout, "That's not fair!" if you felt slighted? Well, AI tools need to ensure they're fair to all students. It means not favoring one group over another and being aware of any biases that might creep in. Because let's face it, no one likes playing a game where the rules aren't the same for everyone.

Lastly, there is accountability. If something goes awry (and let's be real, sometimes things do), there should be a clear way to address it. Whether it's a glitch in the system or an unintended consequence, there should be mechanisms to take responsibility and rectify issues.

KEY CHAPTER TAKEAWAYS

Personalized Learning Defined: Personalized learning is about tailoring education to fit students' unique needs, strengths and interests, ensuring they are engaged and empowered in their educational journey.

The Power of AI in Education: AI-driven tools, like adaptive learning platforms, harness student data to provide a more tailored and responsive learning experience. However, with this power comes the responsibility of ensuring data privacy and addressing algorithmic biases.

Ethical Considerations: As AI becomes more integrated into education, it's crucial to address concerns related to data privacy, algorithmic bias and equity to ensure that all students benefit fairly from these advancements.

CHAPTER LEARNING ACTIVITIES

ACTIVITY 1: PERSONALIZED LEARNING IN ACTION

Reflect on when your learning was truly personalized. What made it so? How did it differ from other learning experiences? Write a short essay or journal entry about this experience.

ACTIVITY 2: DATA AND YOU

Think about the digital footprints you leave behind daily. List down five types of data you think online platforms might collect about you. How do you feel about this data being used to tailor your experiences, whether in learning or other areas of life?

ACTIVITY 3: DEBATE ON ETHICS

Organize a debate or discussion group on the topic: "Is it ethical to use AI in education, considering the potential risks and benefits?" Prepare arguments for both sides, even if you have a strong opinion one way or the other. This exercise will help you see the nuances of the issue.

3
ENHANCING TEACHING WITH AI

THE ROLE OF AI IN LESSON PLANNING, AUTOMATED CURRICULUM DESIGN AND CONTENT SUGGESTIONS

Imagine it is Sunday evening, and instead of spending hours crafting the perfect lesson plan for the week, you're curled up with a good book or maybe even catching up on that TV show you love. Sounds too good to be true? With AI stepping into lesson planning, this dream can be a reality.

Now, I know what you're thinking. "AI? In my classroom? But I'm the teacher!" And you're right. You are the heart and soul of the classroom but stop and think how AI can enhance your capabilities.

With its vast computational power, AI can analyze curriculum standards, understand student needs and even keep track of the latest educational trends. It can then suggest content that's relevant and engaging for your students. Think of it as having a personal assistant who read every book, watched every educational video and even attended every teaching workshop.

AI can also help in designing a flexible curriculum. Remember that one time you had to shuffle around lessons because of an unexpected event? With AI, adapting to changes becomes smoother. It can rearrange topics, ensuring prerequisites are still met and learning remains coherent.

I once tried baking without a recipe, throwing in a bit of this and a dash of that. The result was awkward to say the least. Lesson planning without a clear structure can sometimes feel the same way but with AI's assistance, it is like having a master chef guiding you, ensuring the ingredients will come together just right.

SMART CONTENT CREATION

AI-DRIVEN TOOLS FOR CREATING INTERACTIVE CONTENT AND MULTIMEDIA

Smart content creation through AI-driven tools is revolutionizing the educational landscape, making it easy for educators to create immersive and interactive learning experiences.

Traditionally, creating engaging educational content has been a time-consuming process, often limited by the resources available and the technical skills of the educator. Teachers would spend hours curating and crafting lesson plans, worksheets and presentations, yet these materials might not fully capture the attention or meet the diverse learning needs of their students.

AI-driven tools for content creation are like having a digital assistant who's both incredibly skilled in technology and deeply knowledgeable about your subject matter.

Examples of AI-Driven Content Creation

Virtual Reality (VR) and Augmented Reality (AR): AI can help educators create VR and AR experiences that allow students to explore historical sites, dissect virtual frogs or visualize complex scientific phenomena. This makes learning more engaging and helps students understand abstract concepts by visualizing them in three dimensions.

Interactive Simulations: AI can generate simulations for subjects like physics, chemistry or biology, where students can manipulate variables and see the outcomes in real-time. This interactive approach can demystify complex theories and enhance conceptual understanding.

Adaptive Learning Materials: AI can create adaptive learning paths for students, where the content adjusts in complexity based on the learner's performance. This ensures that each student is challenged appropriately and can progress at their own pace.

Natural Language Processing (NLP): AI tools equipped with NLP can help create easy-to-understand and engaging content. They can summarize extensive information into bite-sized, digestible pieces, generate quizzes and even provide feedback on written assignments.

User-Friendly Interfaces: One of the most significant advantages of these AI tools is their user-friendly nature. Educators do not need to have extensive technical knowledge to use them. With intuitive interfaces and guided processes, teachers can incorporate multimedia elements, interactive content, and personalized learning experiences into their lessons with just a few clicks.

AI MONITORING

TRACKING STUDENT BEHAVIOR, ATTENTION AND PARTICIPATION

Remember when you wished you had eyes on the back of your head? Or when you tried to decipher if Jamie in the back row was taking notes or perfecting his doodling skills? The future has heard our silent pleas, bringing AI into the mix!

Before you jump to conclusions, I'm not suggesting we turn our classrooms into something out of a dystopian novel. No, no. I am pointing us to the fact that AI tools can give us insights into which students are truly engaged and which might be drifting into daydream land. Think of these tools as your classroom's "magic mirror"—offering a glimpse into the unseen.

AI-driven classroom management tools can monitor student participation in real-time. They can provide feedback on who participates in discussions and who might need a nudge. And for those concerned about "Big Brother" vibes, rest assured, it's not about surveillance but support.

Remember when the only tech in the classroom was that bulky overhead projector? Fast forward to today and we've got AI helping us ensure every student receives the attention they need. It's like having a co-teacher who's always on the ball.

AI AUTOMATION OPPORTUNITIES

ADMINISTRATIVE TASKS, ATTENDANCE AND SCHEDULING

I'm talking about those pesky duties that, while essential, can sometimes feel like they're eating up all our time. Remember the days when you'd fumble with paper attendance sheets, trying to decipher handwriting that looked suspiciously like hieroglyphics? Or those moments when scheduling felt like trying to solve a Rubik's cube blindfolded? Well, those days might just be numbered.

Imagine walking into your classroom, and as you start your lesson, a system discreetly checks in each student using facial recognition or voice prompts. No more roll calls, no more interruptions. Think of it as your classroom's assistant, always ready, efficient and never needing a coffee break!

And let's not even get started on scheduling. With AI, it's like having a personal secretary that can optimize timetables, ensure no two exams clash and even reschedule classes when unforeseen events arise.

TEACHER PROFESSIONAL DEVELOPMENT

CONTINUOUS LEARNING WITH AI: PLATFORMS OFFERING PERSONALIZED PROFESSIONAL DEVELOPMENT PATHS FOR TEACHERS

Continuous learning with AI-driven platforms is a transformative approach to professional development for teachers. Traditional professional development often involves generic workshops or online courses that may not fully meet individual teachers' unique needs and challenges.

AI-powered platforms, on the other hand, are designed to understand each teacher as a distinct individual. They consider your strengths, areas for improvement, career goals and even your teaching style. This personalized approach is like having a personal trainer who tailors your workouts to your fitness level and goals.

These platforms suggest courses, workshops and resources tailored to your professional growth needs. If you want to master a new teaching strategy, delve into educational technology or explore classroom management techniques, the platform provides a curated selection that aligns with your aspirations.

The value of these platforms extends beyond content recommendations. They offer valuable feedback on your progress and even help you track your growth over time. Furthermore, they facilitate connections with a community of like-minded educators. This collaborative aspect is essential as it provides opportunities for peer learning, sharing experiences and gaining insights from others who may have encountered similar challenges.

AI-driven continuous learning for teachers enhances their professional growth and significantly impacts the quality of education. Students also benefit when teachers are supported in their ongoing development and access to resources that directly benefit their classrooms. It's a cycle of improvement that can lead to more engaging and effective teaching, ultimately enhancing the learning experience for students.

AI IN PEER REVIEW AND FEEDBACK

Remember those nerve-wracking moments when a colleague would sit in on your class, scribbling notes, and you'd anxiously await their feedback? Before you jump to conclusions and imagine a robot teacher taking over your job, let me clarify. AI isn't here to replace us. Instead, it is like that one friend who gives you honest advice but with a sprinkle of tech magic.

Picture teaching a lesson, and instead of just relying on human observations, there's an AI system quietly analyzing your methods. It's not judging, mind you, it is observing patterns, student reactions and engagement levels. And rather than waiting until your lesson is over, it provides real-time feedback. Gone are the days of waiting anxiously for a colleague's notes. With AI, it's instant and precise.

Now, I know what you're thinking. "But teaching is an art! Can a machine understand the nuances?" Fair point. But think of AI as a magnifying glass. It zooms in on details we might miss. For instance, it might notice that students tend to lose focus fifteen minutes into your lecture or that a particular teaching strategy sparks more engagement than another. The best part is that it can then offer suggestions for improvement—guiding you to be the best version of yourself. Whether tweaking a lesson plan, incorporating more interactive elements or adjusting your teaching pace, AI provides data-based insights.

That said, it is crucial to understand that AI doesn't diminish the human touch but enhances it. It equips us with tools to connect better, teach more effectively and grow continuously. Isn't that what we all want? To be better educators and make a lasting impact?

TRAINING TEACHERS TO INTEGRATE AI IN THEIR TEACHING

Visualize a sunny afternoon—you are in your classroom, prepping for tomorrow's lesson. Suddenly, your computer pings with a notification. It's an AI tool suggesting a new interactive activity tailored to your students' learning styles. Sounds like science fiction? Well, welcome to teaching in the digital age!

Now, I get it. The mere mention of "AI" might have some of you breaking into a cold sweat. "I'm a teacher, not a tech guru!" you might be thinking. Trust me, I've been there. But integrating AI into our teaching isn't as daunting as it sounds. In fact, with the right training, it can be as easy as pie (and who doesn't love pie?).

Remember when interactive whiteboards first hit the scene? There was a learning curve, sure. Fast forward to today however, and they are a staple in most classrooms. The same goes for AI. It's just another tool in our ever-expanding teaching toolkit. But, like any tool, it's most effective when we know how to use it.

That's where upskilling comes into play. It's not about turning us into tech wizards overnight. Instead, it's about equipping us with the skills to harness the power of AI to enhance our teaching. Think of it as a bridge between traditional teaching methods and the digital future. The best part is we don't have to walk that bridge alone. Training programs are popping up left and right, designed specifically for educators like us. A wealth of resources is at our fingertips, from online courses to hands-on workshops and the focus isn't just on the techie stuff. It's about pedagogy, classroom management and, most importantly, ensuring that AI serves our students' best interests.

So, if you're feeling overwhelmed, take a deep breath. The digital age of teaching is an exciting frontier filled with possibilities. As we upskill and adapt, we're not just keeping up with the times—we are leading the charge, shaping the future of education one AI-enhanced lesson at a time.

HUMAN-AI COLLABORATION

You know, there's this common misconception floating around that AI is here to take our jobs but let me tell you a little story. There was a time when calculators were introduced into classrooms. There was this huge uproar about them replacing the need for mental math but look at us now! Calculators didn't replace us; they became tools that enhanced our teaching.

Similarly, AI isn't here to push us out of the picture. Instead, it's like that trusty sidekick, always ready to lend a hand—offering insights or suggest resources without you having to spend hours looking for fresh ideas. That's AI for you! It is about enhancing our capabilities, not replacing them. AI allows us to be more attentive, personalized and efficient but at the end of the day, it's still us, the human educators, steering the ship, with AI as our first mate.

THE BALANCE OF TECHNOLOGY

HUMAN CONNECTIONS IN AN AI-ENHANCED CLASSROOM

Now, onto another essential aspect. With all this tech talk, it's easy to forget the core of education: human connection. I mean, think back to your favorite teacher. Did the fancy gadgets or the heart-to-heart conversations make them stand out? I'm betting on the latter.

While AI can offer data-driven insights and automate tasks, it can't replace the warmth of a teacher's smile or the comfort of a reassuring word. Our classrooms might be getting smarter, but the human touch remains irreplaceable. It's like baking—while a mixer can combine the ingredients, the baker's love and attention to detail make the cake special.

In our AI-enhanced classrooms, it's crucial to strike a balance. Use AI to handle the nitty-gritty, like grading multiple-choice tests or organizing resources. When it comes to discussions, mentorship and fostering a nurturing environment—that's our domain. It's about blending technology's efficiency with a human touch's empathy and intuition.

HUMAN OVERSIGHT IN AI-DRIVEN EDUCATION

Have you ever watched one of those sci-fi movies where robots take over the world? Dramatic, right? I'm not saying that's where we're headed, but it raises a crucial point. As we integrate AI into our classrooms, we can't just set it and forget it.

Imagine AI as a teenager learning to drive. Sure, they've got the basics down, but would you hand over the keys and say, "Good luck!"? Probably not. You would want to be in the passenger seat, guiding them and ensuring they don't take a wrong turn. That's the role we educators play in AI-driven education. We're the safety net, ensuring that the tech doesn't go astray, especially regarding ethical decisions. This may sound obvious, but let's face it, machines lack the moral compass we humans have honed over millennia.

THE FUTURE OF THE TEACHER'S ROLE

The future of the teacher's role in the AI era is a subject of much discussion and speculation. As AI technologies become more integrated into the educational landscape, it's natural to wonder how this will affect educators.

Historically, educators have always adapted to new technologies. Just as the transition from overhead projectors to digital slides didn't make teachers obsolete, the rise of AI in education isn't about replacing teachers but rather about enhancing their capabilities. Teachers have continually embraced new tools to enrich the learning experience, and AI is another step in this ongoing journey. With AI handling tasks such as grading, personalized content delivery and even some aspects of instruction, teachers' roles will likely shift. Educators will have more capacity to act as mentors and guides. They can focus on facilitating deeper understanding, fostering critical thinking and encouraging problem-solving skills.

One of the most significant shifts will be the increased focus on uniquely human teaching skills. Creativity, emotional intelligence and interpersonal skills are areas where teachers can have a profound impact. These are the skills that machines are far from mastering and are increasingly valued in a world where AI often augments technical abilities.

As the role of information provider is increasingly shared with AI, teachers can take on the role of life coaches. This means they can concentrate on helping students navigate the complexities of life, providing support and guidance on issues beyond academic content.

AI may be able to process information and identify patterns, but it cannot replace the human touch teachers provide. The relationships teachers build with students, the understanding and empathy they offer and the ability to inspire and motivate cannot be replicated by AI.

Professional development will be crucial to prepare for these shifts. Teachers must be trained in using AI tools effectively and how best to leverage their unique human skills in an AI-augmented classroom.

KEY CHAPTER TAKEAWAYS

AI as a Tool, Not a Replacement: AI can assist in lesson planning, content creation, and administrative tasks, but it doesn't replace the need for human intuition, empathy and connection in the classroom.

Professional Development with AI: AI platforms can offer personalized professional development paths for teachers, ensuring continuous growth and adaptation to the latest educational trends.

Human-AI Collaboration: The future of teaching isn't about AI replacing educators but about educators leveraging AI to enhance their teaching methods, ensuring a more personalized and efficient learning experience for students.

CHAPTER LEARNING ACTIVITIES

ACTIVITY 1: AI IN YOUR CLASSROOM

Identify one area of your teaching where you believe AI could provide significant assistance. Research tools or platforms that cater to this need and consider integrating them into your lessons.

ACTIVITY 2: REFLECT ON BALANCE

Write a short reflection on how to balance using AI tools and ensuring human connection in your classroom. What are the potential challenges and how might you overcome them?

ACTIVITY 3: PROFESSIONAL DEVELOPMENT PLAN

Using the concept of AI-driven personalized professional development, outline a plan for your growth over the next year. Identify areas in which you'd like to improve and seek AI platforms or tools to assist in this journey.

4

AI IN ASSESSMENT AND FEEDBACK

MODERN ASSESSMENT METHODS

TRADITIONAL VS. AI-DRIVEN ASSESSMENTS

On one side, we have our trusty traditional assessments. Think multiple-choice questions, short answers and the occasional essay. Such things have been around for ages and for good reason. They're straightforward, easy to administer and give a snapshot of a student's understanding. But, and it's a big but, they come with their fair share of limitations. For one, they're time-consuming to grade and they don't always capture a student's holistic understanding or skills.

Enter AI-driven assessments. Imagine a world where assessments are tailored to each student, adjusting in real-time based on their responses—a system that can instantly grade and provide feedback, highlighting areas of strength and weakness. Amazing, right? That is the power of AI! These assessments can analyze patterns, understand student behavior and even detect when a student might be guessing.

Now, I'm not here to say one is better than the other, both have their place. Traditional assessments offer a tried-and-true method, while AI-driven ones bring innovation and adaptability. It's like comparing a classic novel to an e-book, both tell the story but in different ways.

As educators, we have more tools at our disposal than ever before. We must decide how best to use them, always keeping our students' best interests at heart. As we navigate this brave new world of AI in education, let's remember to balance innovation with the timeless values of teaching. After all, it's all about helping our students succeed.

ADAPTIVE TESTING

HOW AI ADJUSTS QUESTION DIFFICULTY BASED ON STUDENT PERFORMANCE

Have you ever been in a situation where you are trying to find the perfect pair of shoes, and suddenly, the online store magically starts showing you options that fit just right? Now imagine if our students could have a similar experience, but with their assessments. Enter the world of adaptive testing powered by AI.

Let's break down the possibilities of how this works. Picture a student taking a test. They answer the first question correctly and instead of moving to the next pre-set question, the system thinks, "Hmm, let's challenge them a bit more." And voilà! The next question is slightly tougher but what happens if they struggle with that? Then the system goes, "Alright, let's take a step back and try a different angle." Each question adjusting accordingly. It's like having a personal trainer for your brain, constantly pushing and pulling to find the right balance.

The beauty of adaptive testing is that it's, well, adaptive! It tailors the assessment experience to each student's unique abilities. With AI, we're stepping into an era where assessments can flex and mold based on individual performance. Turning assessments into a dynamic exchange, where the steps change based on each user.

Now, I know what you might be thinking. "Is this just another tech gimmick?" No, trust me, it's more than that. It is about ensuring every student gets a fair shot and is challenged just enough to grow but not so much that they're overwhelmed.

AUTOMATED GRADING

Remember those late nights spent grading stacks of papers, sipping on your third cup of coffee and wondering if there's a better way? There is! Let's dive into the fascinating world of automated grading powered by AI.

Imagine a tool that doesn't get tired, doesn't need coffee breaks and can provide instant feedback. Sounds dreamy, right? Here's how it works. At the heart of automated grading is a set of algorithms. Think of them as the recipe behind the magic. These algorithms are trained on vast amounts of data, learning from thousands of essays, answers and even those doodles in the margins. Over time, they learn to recognize patterns, nuances and even the intricacies of human expression.

Amazingly, automated grading is not just about right or wrong answers. These systems can provide feedback on grammar, structure and even the coherence of arguments. It's like having a friendly English teacher and a math whiz rolled into one, living inside your computer.

Now, I'm not saying we should throw out human judgment. Far from it! Automated grading is more like a helpful assistant that takes care of the heavy lifting, allowing us, the educators, to focus on what truly matters: connecting with our students, understanding their needs and guiding them on their learning journey.

BENEFITS AND LIMITATIONS

BENEFITS:

- **Instant Feedback:** AI doesn't need to pull all-nighters. It can grade assignments in a flash, giving students immediate responses. Who doesn't love instant gratification?
- **Consistency is Key:** Humans, bless us, can be a tad inconsistent sometimes. AI, on the other hand, maintains a consistent grading standard, ensuring fairness every time.
- **Personalized Feedback:** With AI, it's not just about the grade but tailored feedback, pinpointing where a student might need extra help.

LIMITATIONS:

- **Lost in Translation:** While AI is smart, it might miss the nuances or the creative flair in a student's answer. It's like explaining why a joke is funny; either you "get it" or you don't!

- **Tech Glitches:** Has your computer ever frozen during a Netflix binge? Yeah, AI systems can have their off days, too.
- **Over-reliance:** There's a danger in seeing AI as the be-all and end-all. We must remember that it's a tool, not a replacement for the human touch.

INSTANT FEEDBACK:

HOW AI PROVIDES REAL-TIME FEEDBACK TO STUDENTS

AI providing real-time feedback to students is a game-changer in education. Traditional feedback in education often involved waiting for a teacher to review assignments or assessments, which could sometimes result in delays and missed learning opportunities. Waiting for feedback could be compared to waiting for your favorite song to play on the radio—a test of patience.

AI systems have transformed this process by offering instant analysis powered by vast databases and advanced algorithms. Whether it's a math problem, an essay or a coding assignment, AI can quickly spot errors, suggest corrections, and offer improvement tips, all in real time.

What sets AI feedback apart is its personalized nature. AI systems can identify general errors and specific issues a student needs to address. For example, in mathematics, AI can identify where a student made a mistake and provide step-by-step guidance on correcting it. In writing, AI can highlight areas where improvements can be made, such as grammar, style or structure. The great thing is, students can access feedback and support whenever needed, whether working on assignments late at night, during weekends or in different time zones.

AI also generates data that can be used to track a student's progress over time. Educators can use this information to identify trends, understand areas where students commonly struggle and adapt teaching methods accordingly. This data-driven approach allows for continuous improvement and optimization of the learning experience.

Real-time feedback provides immediate guidance but also reduces student frustration and promotes a learning/growth mindset. Students can make corrections and improvements immediately, leading to better retention and understanding of the material.

THE IMPORTANCE OF TIMELY FEEDBACK

The importance of timely feedback in education cannot be overstated. Timely feedback catalyzes learning. Much like how immediate feedback helps you adjust and improve when attempting a skateboard trick, it has the same effect in education.

When students receive feedback right after completing an assignment or task, they can quickly recognize their mistakes and understand where they went wrong. This allows them to make adjustments, clarify misunderstandings and grasp concepts more rapidly.

Feedback isn't just about pointing out errors; it's also about guiding students toward a better understanding of the material. When they receive prompt feedback, it provides clarity and context for the learning process. It's like having a GPS navigation system for learning, where students can see the bumps and turns ahead. This helps them stay on course and navigate more easily.

Immediate feedback also empowers students and boosts their confidence. They gain a sense of accomplishment when they receive validation for correct answers and constructive input for incorrect ones. This positive reinforcement encourages them to continue their efforts and persist in their learning journey.

Timely feedback provides a more enjoyable and effective learning experience. It's rewarding to know that you're on the right track and students are more likely to develop a love for learning when they receive consistent guidance and support. It can be compared to the excitement of mastering that skateboard trick—the joy of learning and progressing fuels their enthusiasm for education.

It also encourages students to become self-regulated learners. They learn how to identify their errors and make corrections independently. This self-improvement process is a valuable skill that extends beyond the classroom and into other aspects of life.

DATA-DRIVEN INSIGHTS

Data-driven insights are at the core of AI's ability to provide effective feedback and are a cornerstone of modern educational technology. AI systems are adept at recognizing patterns in large datasets.

In the context of education, these patterns could be common errors in spelling or grammar, frequently misunderstood concepts or widespread issues in students' understanding of theories. AI can identify these patterns by analyzing student work across various subjects and skill levels and provide feedback relevant to the larger student body population needs.

Beyond recognizing general patterns, AI can analyze the work of individual students to provide personalized feedback. AI can detect subtle progressions or recurring difficulties by comparing a student's current work against their past performances. This individual analysis allows AI to offer feedback tailored to each student's unique learning journey.

The goal of AI-driven feedback is to be actionable. It's not enough to tell a student that they've made a mistake; effective feedback explains why it's a mistake and how to correct it. By providing context and clear steps for improvement, AI helps students learn from their errors and make positive strides in their education.

Just as a coach observes an athlete's performance and offers targeted advice to improve their skills, AI observes a student's academic performance and provides specific guidance. This "coaching" is based on a comprehensive analysis of the student's work, ensuring the feedback is relevant and helpful.

With data-driven insights, AI helps students focus on specific areas that need improvement. Instead of a scattergun approach to studying, students can target their efforts where they're most needed, leading to more efficient and effective learning.

The more data AI systems have, the better they become at providing feedback. They learn from the vast array of student interactions, continuously improving their algorithms to offer more precise and helpful insights.

While AI provides data-driven insights, the human element remains essential. Teachers can use AI feedback to better understand their students' needs and inform their teaching strategies. This partnership between AI insights and human expertise truly lights the path forward for students.

MOTIVATING LEARNERS THROUGH DATA-DRIVEN INSIGHTS

Remember that feeling when you couldn't grasp a concept, no matter how hard you tried? Frustrating, right? Now, imagine if someone had told you precisely where you were going wrong and what you could do to improve. That would've been a game-changer!

Enter the world of data-driven insights. Every student has unique strengths, weaknesses and learning curves. By harnessing the power of AI and data, we can pinpoint exactly where a learner might be struggling or excelling. This is not just about identifying these areas but about using this information to motivate and guide.

Imagine telling a student, "Hey, I noticed you're fantastic at solving algebraic equations but seem to struggle with geometry. How about we focus on that?" It's like giving them a roadmap to success. And when students see their progress, understand their journey and recognize their achievements, it boosts their confidence and motivation.

This is not a data thing—it is about tapping into one's humanity and fostering the belief that every student can succeed. Data-driven insights are just tools in our vast educational toolkit that help us do that. They help us see the bigger image, but it's our job, as educators and mentors, to paint it for our students. And trust me, the results are nothing short of magical when we do!

FAIR AND INCLUSIVE ASSESSMENT

THE CHALLENGE OF BIAS IN AI

We all remember those old-school multiple-choice tests where you'd sometimes guess the answers. Ah, the good old days! But what if the test had biases, favoring one group of students over another?

Now, fast forward to today. We've got AI-driven assessments, which, at first glance, seem like the epitome of fairness. Machines don't have feelings, so they can't be biased, right? Well, not quite. Here's where things get a tad complicated.

AI, in all its glory, learns from data. This means, if that data carries biases from the real world (which, let's face it, our world is riddled with), the AI will, unfortunately, pick up on those biases. It's like teaching a parrot to talk; if you only teach it pirate phrases, don't be surprised if it crows, "Ahoy, matey!"

But why should we care? Because biased AI assessments can perpetuate existing inequalities in education. Imagine a brilliant student not getting the recognition they deserve because of an underlying bias in the system.

Now, while AI has its challenges, it also reflects us humans. It's our responsibility to ensure that the data we feed into these systems is as unbiased and all-inclusive as possible. It's a tall order, but if we can put a man on the moon, we can tackle this, right?

STRATEGIES FOR ENSURING FAIRNESS

You've got your ingredients, your recipe and your enthusiasm but if one of those ingredients is off, your cake might not turn out as you'd hoped. Similarly, with AI, the outcome can be less than perfect if the data is biased but fear not! Just as there are ways to save a sinking cake, there are strategies to ensure fairness in AI-driven grading and feedback.

First off, diverse data. Think of it as adding a mix of ingredients to your cake. The more diverse our data, the richer and more inclusive our AI outcomes. Pulling data from various sources and demographics can reduce the chances of any single bias dominating.

Ever heard of the saying, "Too many cooks spoil the broth"? In the world of AI, it's the opposite. Collaboration is key. We can ensure a holistic approach to fairness by involving experts from various fields. It's like getting advice from a baker and a nutritionist when making that cake.

Another strategy? Regular audits. As you occasionally check your cake in the oven, we must regularly review and refine our AI systems. This helps identify any biases that might have crept in and rectify them before they become a major issue.

Lastly, transparency. Let's be open about how our AI systems work. By being transparent about our methodologies and intentions, we can build trust and ensure everyone's on the same page.

CATERING TO DIVERSE STUDENT POPULATIONS

Picture a classroom filled with students. No two faces are the same, and neither are their learning styles, backgrounds or experiences. It's a beautiful mosaic of diversity. Now, imagine an AI system that treats them all the same. Doesn't quite fit the image.

When I think of AI assessments, I often liken them to shoes. Bear with me on this tangent. Just as one shoe size doesn't fit all, a single AI assessment approach won't cater to every student's unique needs. So, how do we ensure our AI assessments are as diverse as our students?

Firstly, understanding the audience is crucial. We must recognize that students come from various backgrounds, cultures and abilities. By tailoring AI assessments to consider these factors, we're not just being inclusive, we are being fair.

Now, here's a thought. Remember those choose-your-own-adventure books? The ones where you decide the story's path? What if our AI assessments were a bit like that? Multiple assessment pathways based on a student's learning style or background can make a difference. Visual learner? There's a path for that. Prefer auditory cues? There's a route for you, too.

Lastly, let's talk about feedback. It's one thing to assess, but it's another to provide constructive feedback that resonates. By ensuring our AI systems give culturally sensitive and personalized feedback, we're not just grading but guiding.

THE MORAL IMPLICATIONS OF AI IN STUDENT ASSESSMENT

AI systems excel at efficiently processing vast amounts of data. However, AI falls short when truly understanding the depth of human experiences, perspectives and cultural nuances. For instance, grading an essay like Mia's, which draws on personal experiences and cultural context, poses a significant ethical challenge.

AI might process the text but struggle to comprehend Mia's perspective genuinely. This distinction between processing and understanding is a fundamental ethical consideration. It emphasizes the importance of human expertise in assessing the richness of human expression.

AI-driven assessments gather valuable data on student performance. However, important ethical questions surround our data collection, storage and usage. Students' data privacy is a substantial concern, particularly involving minors. Understanding where this data goes, who has access to it and how it's safeguarded is critical. The ethical responsibility lies in treating this data with the utmost care, recognizing its sensitivity and the potential consequences of mishandling it.

Bias in AI assessments is a critical ethical issue. It's not about personal preferences but rather the potential for AI to inadvertently favor or discriminate against students based on their backgrounds, writing styles or other characteristics. Ensuring that AI assessments are free from bias is not just a technical challenge but a moral obligation. Assessments must be fair, transparent and equitable, regardless of a student's identity or circumstances.

There's an ethical requirement for transparency in how AI assessments work. Students and educators should understand the criteria and processes used for grading and evaluation. Additionally, there should be mechanisms in place to address discrepancies or concerns. The accountability of AI systems and the organizations implementing them is paramount.

A potential solution to these moral concerns is a partnership between human educators and AI. Combining the strengths of AI, such as rapid processing and data analysis, with the empathetic understanding, contextual appreciation and ethical judgment of humans can address many of these ethical challenges.

The ethical use of AI in education should be part of the curriculum. Educating students about AI's implications and ethical considerations in assessments is vital to raising awareness and promoting responsible AI use. Ethical considerations should not be an afterthought but an ongoing process. Regular assessments of AI systems for fairness, bias and privacy are necessary to ensure proper use.

KEY CHAPTER TAKEAWAYS

Adaptive Testing: AI's ability to adjust the difficulty of questions in real-time offers a personalized assessment that can challenge and support students effectively.

Automated Grading: The use of AI for grading not only saves time but also provides consistent and immediate feedback, allowing for a more dynamic learning process.

Data-Driven Insights: AI's analysis of student performance data can lead to personalized feedback and targeted support, enhancing the learning experience.

Ethical Considerations: Addressing bias, ensuring fairness and maintaining privacy in AI assessments are paramount to upholding the integrity of education.

CHAPTER LEARNING ACTIVITIES

ACTIVITY 1: AI SIMULATION EXERCISE

- **Objective:** To understand the mechanics behind AI-driven assessments.
- **Activity:** Create a flowchart outlining an AI system's decision-making process during an adaptive test. Include pathways for correct and incorrect answers and how the AI would adjust the difficulty level of subsequent questions.

ACTIVITY 2: BIAS DETECTION WORKSHOP

- **Objective:** To recognize and address potential biases in AI systems.
- **Activity:** Review a set of data or case studies using AI in assessments. Identify any potential biases in the data and discuss strategies for mitigating these biases in future applications.

ACTIVITY 3: FEEDBACK ANALYSIS TASK

- **Objective:** To appreciate the depth and personalization of AI-generated feedback.
- **Activity:** Compare AI-generated and human-generated feedback on a sample student's work. Analyze the differences in tone, specificity and action ability. Discuss how each can complement and be integrated in a real-world setting.

5
AI-POWERED EDUCATIONAL CONTENT

CONTENT CREATION AND CURATION

UNDERSTANDING THE CAPABILITIES OF AI IN CONTENT GENERATION

Remember when we used to handwrite essays? Or when research meant hours in a library, flipping through dusty books? Fast forward to today, and we're in an era where AI can whip up content in a jiffy. Mind-blowing, right?

AI isn't just a fancy robot that spits out words. It's like a chef blending ingredients (data) to cook a delicious dish (content) but instead of following a recipe, it learns from a variety of vast information. It's like reading every book on a subject ever written and then writing an essay. Talk about superhuman!

Some folks ask, "Can AI capture the human touch? The emotions? The nuances?" Great questions! While AI is a champ at processing data and spotting patterns, it doesn't "feel" emotions like we do. Think of it like a super-efficient assistant who's all logic and no drama. It's handy, but don't dare ask for relationship advice!

Now, imagine AI as a budding artist. At first, it mimics famous painters but with enough data and learning, it starts creating its own unique masterpieces. That's AI in content generation. It's not just regurgitating, it is creating.

But, and this is a big BUT, while AI is a game-changer, it's not the be-all and end-all. It's a powerful tool, but it still needs us, the humans, with all our creativity, our experiences and our stories. So, while AI makes waves in content generation, we must ride those waves and steer the ship.

HOW AI CAN REDUCE THE BURDEN OF CONTENT CREATION

I remember my days in the classroom, juggling lesson plans, grading and trying to keep up with the latest educational content. It felt like I was in a never-ending marathon, where the finish line kept moving further away. Sound familiar? As teachers, I think most of us can relate.

Fast forward to Sunday evening, and you're prepping for the week. Instead of spending hours searching for relevant content, why not let an AI tool scan the vast internet, curate the best resources and suggest ways to integrate them into your lessons. It's like having a personal librarian and curriculum planner all rolled into one.

Now, let's talk about customization. We already agreed that the one-size-fits-all doesn't work in education but to write thirty or fifty variations of each lesson for each student is humanly impossible. Here's where AI shines. It can analyze individual student data and tailor content to fit their learning style and pace. Think of it as a bespoke suit for each student. And the best part? You don't have to do the heavy lifting.

But wait, there's more! Remember when you wanted to introduce a new teaching method or a fresh topic but didn't have the resources? AI can generate content on the fly. It's like having a creative genie that grants educational wishes.

Now, I know what you're thinking. "This all sounds great, but what about my personal touch?" And you're spot on. Once again, AI is a tool, not a replacement. It's there to assist, not take over. It's like when you use a dishwasher. It does the cleaning, but you decide how to arrange the plates and which detergent to use. The control is still in your hands.

ENSURING THE ACCURACY AND RELEVANCE OF CONTENT

So, you've heard about AI-generating content, and if you're anything like me, you have probably thought, "Sounds cool, but can I trust it?" It's like when you first hear about self-driving cars. The idea is fascinating, but there's that nagging question: "Is it safe?" Well, let's unpack this together.

First off, let's get something straight. AI is impressive, but it's not infallible. Think of it as a super-smart student in your class. They're brilliant but still need guidance, checks and balances. That's where quality control comes into play.

Remember that 'AI chef' we have been talking about? We have all agreed that it can probably whip up a dish (or content) based on the ingredients (data) given but what if the ingredients are stale or, worse, contaminated? The dish won't be up to the mark. Similarly, the data we feed into AI determines the quality of its content. Garbage in, garbage out, as they say. So, what is the first step in quality control? Feed it good, clean and relevant data.

But there's more to the story. Once AI dishes out the content, we need a taste test and a review process. This is where human expertise shines. Educators can review, tweak and ensure the content aligns with our curriculum and educational standards. It's a partnership. AI brings the efficiency and we bring the discernment.

Now, let's talk about relevance. Have you ever had a student share an interesting fact that's not quite related to the topic at hand? AI can sometimes do that, too. It might generate content that's factually correct but not contextually relevant. This is where continuous feedback loops come in to play. By providing feedback, we train the AI to understand the context better, ensuring the content it produces is accurate and pertinent.

CUSTOMIZED LEARNING MATERIALS

WHY CUSTOMIZATION MATTERS IN MODERN EDUCATION

Have you ever tried on a one-size-fits-all shirt? If you're anything like me, you've found that it doesn't quite fit right. It might be too baggy in some places and too tight in others. Now, let's think about this in the context of education.

You see, just as we all have unique body shapes and sizes, our minds are wonderfully diverse, too. Each student comes with their own set of experiences, interests, strengths and challenges.

So, why would we expect a single teaching method or material to resonate with everyone? It's like expecting everyone to enjoy the same flavor of ice cream. (And between you and me, while I'm a die-hard chocolate fan, I know some folks who'd pick mint choc chip any day!)

Remember when mix tapes were a thing? Crafting a playlist tailored to someone's taste? Personalization in education is a bit like that. It's about curating a learning experience that resonates with each student, making them feel seen and understood.

The traditional approach to education is becoming outdated in today's fast-paced, ever-changing world. We're in the age of choice where Netflix, Spotify and Amazon algorithms curate content based on individual preferences. Shouldn't our educational content be just as tailored?

Personalized learning materials can address individual needs, cater to different learning styles and ignite a passion for learning. When students see content that reflects their interests and experiences, they're more likely to engage, participate, and, most importantly, learn.

HOW AI TAILORS AND ADAPTS CONTENT

AI tailoring content to individual learning styles and needs is akin to a highly personalized service experience. In a traditional classroom, teachers strive to meet individual student needs, but it's a challenging task due to time constraints and the sheer number of students. AI, however, can personalize learning at scale, simultaneously provide customized content for each student. AI systems can dynamically adapt learning materials based on real-time feedback from student interactions. If a student excels in a subject, the AI can introduce more advanced concepts or accelerate their pace. Conversely, if a student is struggling, the AI can slow down, revisit foundational concepts, or present the information in a different format that may be easier for the student to understand.

Engagement is crucial for effective learning. AI can enhance engagement by aligning educational content with each student's interests. For example, suppose a student shows a keen interest in space. In that case, the AI can integrate space-related examples and content into math or physics problems to keep the students engaged and motivated.

Different students have different learning preferences—some learn better through visuals, others through auditory materials and others through kinesthetic experiences. AI can cater to these preferences by presenting content in various formats, such as videos, interactive simulations, or text, depending on what works best for them.

AI systems are great at creating a continuous loop of learning, assessment, feedback and adjustment. They assess a student's performance on various tasks, provide immediate feedback and then adjust the subsequent content to better suit their learning needs.

AI can uncover insights into learning patterns and preferences by analyzing how students interact with content. This data can inform the creation of new content that is even more closely aligned with student needs.

Advanced AI systems can also consider the student's emotional and cognitive state. For example, suppose a student consistently struggles and shows signs of frustration. In that case, the AI can adapt the difficulty level or provide encouraging feedback to help maintain the student's confidence.

As AI systems collect and analyze student data to personalize learning, we must remember the importance of handling this data carefully. This includes securing data against unauthorized access and ensuring that the personalization algorithms are transparent and free from bias.

HOW STUDENT DATA INFORMS CONTENT PERSONALIZATION

I'm not talking about those endless spreadsheets we sometimes dread. No, I'm diving into how student data is revolutionizing content personalization. Imagine it's a rainy afternoon, and you're binge-watching your favorite show. Suddenly, you get a recommendation for another series just up your alley. How did they know? It's all about the data you've unknowingly provided through your viewing habits. Now, let's bring that concept into our classrooms.

Every time students interact with a digital platform, whether by answering a quiz, watching a video or even spending time reading about a particular topic, they leave behind digital footprints. These aren't just random bits of information but valuable insights into their learning styles, interests, strengths and areas that need more attention.

AI systems, being the clever things they are, analyze this data and tailor content to fit each student's unique needs. It's like having a personal tutor for every student, ensuring they get the right content at the right time. Imagine the boost in confidence when a student struggling with a concept gets resources tailored just for them!

While data is powerful, it's essential to use it responsibly. After all, we are shaping young minds, and that's a responsibility we shouldn't take lightly.

ADDRESSING POTENTIAL PITFALLS IN AI-DRIVEN CONTENT

We've chatted about the wonders of AI in education, but like that one marshmallow that always ends up a bit too toasted, there are some challenges we need to address.

First, let's talk about the elephant in the room: Data privacy. We're collecting heaps of student data to tailor their learning experiences, but where's that data going? Who has access? And for how long? It's crucial to ensure that student information is protected and not misused. Remember, with great data comes great responsibility.

Now, here's a curve ball for you. Ever heard of the term 'overfitting'? In the AI world, it's when the system gets too good at customizing content based on past data, making it less adaptable to new information or different learning scenarios. It's like when I tried to use my grandma's secret cookie recipe for muffins. Didn't quite work out.

And let's not forget about potential biases. If the data fed into AI systems isn't diverse, its content might inadvertently favor one group over another. It's like teaching a parrot only to sing show tunes and then being surprised when it doesn't know any classic rock.

Lastly, there's the risk of over-reliance. While AI can be a fantastic tool, we must strike a balance, ensuring that technology enhances, not overshadows, the personal connections that make learning truly impactful.

CREATIVE APPLICATIONS OF AI

Have you ever daydreamed about taking your students to Mars or diving deep into the human cell? With AI, we're not just talking about reading from a textbook. We're venturing into virtual labs, simulations and interactive scenarios.

Instead of explaining a chemical reaction, students can experience it in a virtual lab, adjusting variables and seeing the results in real-time. Or how about history lessons where students can virtually walk through ancient civilizations, interacting with historical figures? It's like turning the classroom into a time machine!

And for those hands-on learners, AI-driven simulations can mimic real-world scenarios. Think of it as a digital sandbox where students can experiment, make mistakes, and learn without real-world consequences. It's learning by doing, supercharged!

But wait, there's more! Have you ever thought about creating interactive stories where students choose their own adventures? With AI, these narratives can adapt to individual choices, ensuring a unique learning journey for each student. It's storytelling with a twist!

AI IN ARTS AND CREATIVITY

For all art teachers, AI tools enable students to create masterpieces, blending traditional techniques with digital magic. Imagine a student sketches a simple landscape, and AI helps transform it into a detailed painting, suggesting color palettes and brush strokes. What about our budding musicians, you might ask? AI can analyze their compositions, offer suggestions or even create harmonies. It's like having a virtual music buddy jamming alongside you!

But here's where it gets wild. Literature enthusiasts, imagine an AI tool that helps students craft stories, offering plot suggestions, character developments, or even poetic verses. It's not about replacing the human writer but enhancing it, making the creative process even more enriching.

GAMIFICATION AND AI

Ever noticed how students are so engrossed in games? Now, what if we could channel that same energy into learning? Enter the world of gamification powered by AI. Take a math lesson for example. With a few clicks, AI can transform that 'once boring problem,' into a quest where students solve problems to unlock new levels. Or a history class where they embark on time-traveling adventures. With AI, these games can adapt to each student's pace and learning style. Struggling with a concept? The game offers hints. Mastered a topic? It ups the challenge. It's like having a personalized video game for every subject! While students think they're just playing, they are actually learning; and not just the curriculum but also critical thinking, problem-solving and teamwork.

FUTURE PROSPECTS

With the pace at which AI is evolving, what might the classrooms of 2030 look like? Imagine a Virtual AI teacher co-existing with human educators, each bringing their strengths to the table or students collaborating on projects in virtual reality spaces, guided by AI mentors. How about AI tools that can craft personalized learning journeys, predicting which topics a student might struggle with weeks in advance?

For the creatives, image AI-powered tools that help students compose music, write stories or create art all tailored to their unique style and preferences. The possibilities are endless! AI will undoubtedly play a massive role but I can't say this enough, and I will keep repeating it until this sinks—the human component will always be irreplaceable. Education is not complete without the passion and empathy human educators bring to the table. So, as we ride this wave of innovation, let's remember to balance tech with touch, ensuring that education remains a deeply human experience.

KEY CHAPTER TAKEAWAYS

- AI has the potential to revolutionize content creation and curation by providing personalized, engaging and multi-modal learning experiences.
- Data privacy, ethical use and avoiding over-reliance on technology are crucial considerations in integrating AI into education.

- The future of education with AI promises even greater personalization, but the irreplaceable value of human interaction and empathy must be preserved.

CHAPTER LEARNING ACTIVITIES

ACTIVITY 1: DATA DIVE AND DESIGN

- **Objective:** Understand the importance of data in personalizing education.
- **Activity:** Create a mock "data profile" for a fictional student based on their learning habits and preferences. Then, design a week of personalized learning activities using AI recommendations tailored to this profile. Reflect on how this personalization might impact the student's engagement and learning outcomes.

ACTIVITY 2: ETHICAL ROUNDTABLE

- **Objective:** Explore the ethical considerations of AI in education.
- **Activity:** Organize a discussion group or roundtable to debate the ethical implications of using AI in educational settings. Topics can include data privacy, bias in AI algorithms and the balance between technology and human teaching. Summarize the key points of consensus and contention and propose guidelines for ethical AI use in education.

ACTIVITY 3: CREATIVE AI WORKSHOP

- **Objective:** Experience the intersection of AI and creativity.
- **Activity:** Use an AI-powered tool to create a piece of art, music or a short story. Reflect on the process, noting how the AI contributed to the creative outcome. What were the benefits and limitations? Share your creation and insights in a classroom or workshop setting to spark a conversation about the role of AI in the creative process.

6
THE FUTURE OF AI IN EDUCATION

EMERGING TRENDS AND TECHNOLOGIES

PREDICTIONS FOR THE FUTURE OF AI IN EDUCATION

As someone deeply entrenched in the world of education, I often wonder, "What's next for AI?" We've seen some pretty cool stuff, but what does the future hold?

Remember when AI was all about recognizing patterns and making predictions? Well, the next wave is going to be all about understanding. We're talking AI that doesn't just process information but truly grasps context, emotion and nuance. Think of an AI that can engage in meaningful, deep conversations about literature, philosophy or even the latest pop culture trends.

We're on the brink of seeing AI that can learn and adapt without being explicitly programmed. They can learn new subjects or teaching methods simply by observing human educators.

Think about the potential for global classrooms. With advanced AI-driven translation tools, students worldwide could learn together in real time, breaking down language barriers and fostering global collaboration and understanding.

Now, I know what you're thinking. "This sounds amazing, but what does it mean for us educators?" Well, it's simple. As AI takes on more administrative and routine tasks, we focus need to focus on inspiring and mentoring—things that no machine can replicate.

INTEGRATION WITH OTHER TECHNOLOGIES

With its incredible data processing capabilities, AI meets quantum computing, the new kid on the block that promises to solve complex problems in the blink of an eye. It's like pairing peanut butter with jelly or scones with clotted cream for my fellow Brits.

Quantum computing, with its qubits and superposition, could potentially process vast amounts of data at speeds we've never seen before. Now, combine that with AI's ability to learn and adapt. The result is an ultra-smart system that can predict, analyze and respond to just about any educational challenge that comes its way. Imagine a virtual classroom that adapts instantly to classroom lessons or research projects that would've taken years, completed in mere days. Amazing, right?

THE ROLE OF BIG DATA

We have seen that big data offers unprecedented insight into the learning process but think of what it can do for long-term learning. Big data allows predictive analytics to forecast future learning outcomes based on historical data and plan a long term course for each student to reach their desired outcomes.

Such data can even help develop a curriculum by predicting which subjects or skills will be most valuable in the future job market.

Educators can make informed decisions by analyzing big data and improving educational outcomes. For instance, data analysis can reveal the most effective teaching strategies or the optimal time to introduce certain subjects, leading to better student engagement and understanding.

Big data can highlight educational opportunities and gaps in outcomes. Educators can identify disparities and work towards more equitable education systems by examining data across demographics. It is the job of educators and institutions to ensure that students benefit from big data, rather than be harmed it. Just like traditional school, planned for infrastructure measures to keep students safe (think fences and gates), institutions of the future will need to invest in robust cybersecurity measures to protect against data breaches and unauthorized access.

Transparency in collecting, using, and sharing student data is also key. Students and parents should be informed about what data is being collected and how it will be used. Obtaining consent is a critical component of ethical data use and should be part of a children's school registration process.

PREPARING FOR THE FUTURE

HOW TO ADAPT TO AN AI-DRIVEN LANDSCAPE

I often wander down memory lane, reminiscing about when "technology in the classroom" meant having a single computer with that nostalgic Windows 95 startup sound. The times when a projector was the coolest thing and clunky VHS tapes were our window to the outside world. Ah, those were simpler times, but wait, am I showing my age?

Fast forward to today, and here we are, standing on the brink of what seems like a sci-fi reality. AI-driven tools, virtual classrooms and students coding better than they can write essays. I mean, who would've imagined?

With all this tech jazz, does the role of educators get thrown into the shadows? Not quite. Our roles are not diminishing; they're evolving. Gone are the days when teachers were the primary source of information. With a quick Google search, kids have the world at their fingertips. So, what's our new role? It's to guide these young minds through the vast sea of information, helping them discern fact from fiction.

Let's be real. No AI tool can replicate the magic of human emotion. Remember the joy when your teacher praised you? Or the comfort when they reassured you after a bad day? That's irreplaceable. Our job is to ensure that the human essence remains alive and kicking in this digital world.

The tech world is sprinting, and we must keep pace. Professional development isn't a choice; it's a necessity. Dive into workshops, get certified, join online forums—keep that learning flame burning!

INFRASTRUCTURE AND INVESTMENT

To those hallways adorned with years of memories and those classroom walls that echoed with laughter—good times! But while reminiscing, it struck me; are these age-old infrastructures ready for the digital dawn that's upon us? It is like trying to fit a modern engine into a vintage car. Sounds mismatched, right? Similarly, with their legacy infrastructures, our schools might not be primed for the technological renaissance awaiting them. That's where infrastructure and investment comes into play.

Remember when 'smart' classrooms were the talk of the town? I remember feeling like a tech wizard just by using a smart board! Today, it's not just about a tech gadget here and there. It's about creating an environment where tech integrates seamlessly in all aspects of the classroom. Think virtual reality history lessons, automated attendance and AI-driven curriculum.

Here's a little nugget I picked up at a recent seminar (and yes, there were virtual cookies); with great tech comes great responsibility. Our institutions need robust cybersecurity systems. It's not just about protecting data but about ensuring our kiddos surf the digital waves safely.

Open those purse strings to the policymakers and bigwigs out there! Investing in education infrastructure is not an expense, it's an investment in our future leaders, thinkers and innovators.

THE IMPORTANCE OF CONTINUOUS LEARNING

Remember when we thought our learning days were behind us once we tossed that graduation cap into the air? Reality served me a piping hot slice of humble pie a few years into adulthood. Education doesn't end with a degree, folks. It's a lifelong journey, especially in our pulsating, AI-driven world.

In this rapidly evolving era, what we learned yesterday might become obsolete tomorrow. Wild, right? And the culprit? A cool, sassy invention called Artificial Intelligence. Take Alexa, Siri or any of your virtual buddies. Today, they're setting reminders and playing our favorite songs. Tomorrow? Who knows, maybe they'll teach algebra or give us lessons on Shakespearean plays (I hope they've got a good British accent!). This means two things:

- **Students, Stay Curious:** To our younger generation, remember the AI wave isn't just about machines. It's an invitation. It is an invitation to be adaptable, constantly curious and embrace learning as a dear old friend who's always got new stories to share.
- **Educators, You're in This Too:** For my fellow educators, don't think you're off the hook! Teaching in the AI age also means we need to be perpetual learners. We're no longer just the providers of knowledge but co-learners navigating this brave new world alongside our students.

INTEGRATING AI INTO VARIOUS EDUCATIONAL SETTINGS

Integrating AI isn't about having the fanciest robots or the sleekest software. Nope. It's all about strategy.

Start Simple: Have you heard of the KISS principle? 'Keep It Simple, Silly!' Begin with small AI tools that assist in tasks like grading or scheduling. There is no need for a grand entrance with a robot teacher yet!

Enhance, Don't Replace: Here's a controversial take. Machines can never replace the warmth, empathy and connection human educators bring. Use AI to enhance, not replace.

Stay Updated: Ever bought a smartphone and then, boom, a new model's out? That's tech for you! Always on the move. Keep an eye out for the latest trends and tools. Maybe join a workshop or two? Wink!

Mix human interaction with AI. Maybe a group project where students use AI data analysis or a storytelling session with an AI tool? Fusion is the key.

Get Feedback: Always, and I mean ALWAYS, get feedback from students, teachers and parents. Heck, even the school janitor might have a golden nugget of insight!

Keep Ethics A Priority: With great power...you know the drill. Ensure students know the ethical implications of AI.

KEY CHAPTER TAKEAWAYS

Emerging Trends: AI's future in education is not just about recognizing patterns but truly understanding context, emotion and nuance. The integration of quantum computing and AI promises unparalleled data processing speeds.

Integration with Other Technologies: As AI becomes more integrated with other technologies, the need for robust infrastructure, investment and cybersecurity is paramount.

Continuous Learning: In the AI-driven world, students and educators must embrace continuous learning. The landscape is ever-evolving and adaptability is key.

Best Practices: When integrating AI into educational settings, it's essential to start simple: remember that AI is a tool (not the teacher). Stay updated, promote collaborative learning, gather feedback and always prioritize ethics.

CHAPTER LEARNING ACTIVITIES

ACTIVITY 1: AI TOOL EXPLORATION

Spend a week exploring a new AI tool relevant to your subject or area of interest. It could be a grading tool, a scheduling assistant or even an AI-driven content creator. Document your experiences, noting the benefits and challenges.

ACTIVITY 2: ETHICS DISCUSSION

Organize a group discussion or debate on the ethical implications of AI in classrooms. Topics could include data privacy, potential bias in AI algorithms or the balance between technology and humanity.

ACTIVITY 3: FUTURE CLASSROOM DESIGN

Create a blueprint or a mock-up of what you believe a future AI-enhanced classroom might look like in 2030. Consider infrastructure, tools and the role of the teacher and students in this environment.

7
AI IN SPECIAL EDUCATION

I'm sure many of you have witnessed firsthand the challenges faced by students with disabilities. It's a unique journey that requires patience, understanding, and, above all, tailored approaches. With the rise of AI, we're on the brink of a revolution that could change the game for these incredible learners.

When I think of education, I often imagine it as a vast ocean. While many students sail smoothly, some face turbulent waters due to their unique learning needs but visualize a world where every student has a customized map to navigate these waters regardless of their challenges. That's where AI steps in, like a seasoned captain, guiding the ship.

EXAMPLES OF ASSISTIVE AI TECHNOLOGIES

SPEECH RECOGNITION AND GENERATION

Have you ever tried having a conversation without speaking? It's challenging. Imagine students with hearing or speech impairments trying to communicate their thoughts. Enter AI-driven speech recognition and generation tools. These nifty gadgets can convert spoken words into text in real-time, ensuring that students never miss a beat in class. And for those who struggle with speech? AI can generate human-like voices, allowing them to "speak" their thoughts with ease. With AI, disabilities are a thing of the past!

VISUAL ASSISTANCE

Imagine a student with a visual impairment navigating a textbook with the help of an AI tool that describes images, graphs and even complex diagrams. Sounds futuristic? Well, it's happening right now. AI-driven solutions transform texts into rich auditory experiences, making learning accessible and engaging. From describing intricate art pieces in history books to breaking down complex scientific diagrams, AI ensures no student is left in the dark.

MOBILITY AND PHYSICAL ASSISTANCE

I've got to admit, this one is my favorite. Imagine tiny robots helping students with physical disabilities navigate school corridors or even assisting with writing or turning pages. It's not a scene from a movie; it's the combined power of AI and robotics. These tools are designed to be an extension of the student, ensuring they can participate in classroom activities seamlessly. From robotic arms to AI-powered wheelchairs that can navigate obstacles, the future is here, and it's awe-inspiring.

COGNITIVE SUPPORT

Learning disorders or cognitive impairments can often make traditional learning methods challenging but AI's got our back.

Tailored learning platforms use AI to break down information into digestible chunks, adapting to each student's pace. For a student with ADHD, this might mean interactive lessons with frequent breaks. For another with dyslexia, it could be text displayed in a specific format or color. It's all about understanding the unique needs of each student and crafting a learning experience that resonates with them.

INCLUSIVE LEARNING ENVIRONMENTS

THE PHILOSOPHY OF INCLUSION

Before you think, "Oh, it's just another educational buzzword," let me stop you right there. This isn't just a trend; it's a movement, a belief and a way of life. Here's why.

You know, back in the day, classrooms were pretty uniform. Everyone learned the same way, at the same pace, from the same books but we're not all the same. We come from different backgrounds, have different abilities and see the world through our unique lenses. Inclusion is all about recognizing and celebrating these differences. It's about ensuring that every student, regardless of their abilities, background or challenges, feels valued, heard and seen. It's about creating a space where everyone can thrive, not just survive. Trust me, this is not just some lofty ideal, it's a necessity.

Think about it. What message are we sending when we exclude certain students from the mainstream? That they're not good enough? That they don't belong? Ouch. That's not the kind of world I want to live in. Inclusion is about breaking down these barriers, challenging the status quo and saying, "Hey, every student matters."

Some argue that inclusive education is too challenging or might hold back other students but here's a little food for thought. Inclusion doesn't mean lowering standards, it means adapting them. It's about finding innovative ways to teach so every student can reach their potential. When we raise the bar for one student, we raise it for all. It's a win-win.

Beyond the academics, there's an even bigger image here. Inclusion teaches empathy, understanding and respect. It prepares students for the real world, where diversity is the norm. It fosters a sense of community, where everyone looks for one another and don't we need more of that in today's divided world?

AI-DRIVEN CLASSROOM ADJUSTMENTS

Imagine a classroom that automatically reshapes to fit each student's unique needs. Sounds like a scene from a fantasy novel, right? But what if I told you this isn't just the stuff of fairy tales? Enter the world of AI-driven classroom adjustments. This AI advancement is nothing short of magical!

I remember my days in school, sitting at those one-size-fits-all desks, straining to see the board and sometimes feeling lost in the crowd. Today's classrooms? They're on the brink of a revolution, my friends, and AI is leading the charge.

First off, let's talk about those students who might have visual or hearing impairments. Traditional classrooms can be a challenge but with AI, we're seeing tools that can adjust lighting based on a student's visual needs or optimize acoustics for those with hearing challenges. Imagine a classroom where the lights dim or brighten to suit each student or where sound frequencies are tweaked to ensure everyone hears the lesson. It's like having a personal classroom concierge!

But wait, there's more. AI can adjust the room's physical layout for our kinesthetic learners, those who learn by doing and moving. Desks and chairs that move and rearrange themselves to create open spaces for activities? Yep, that's on the horizon. And for our students who might need a quieter, more focused environment? AI-driven partitions can create instant "focus zones," giving those who need it, their own personalized learning nook.

Now, I hear some of you thinking, "But what about the emotional and cognitive aspects?" AI tools are being developed to sense when a student feels stressed or overwhelmed. Maybe the room's temperature adjusts or calming ambient sounds play subtly in the background. It's all about creating an environment where students feel comfortable and ready to learn.

Remember those sci-fi movies where everything in the room responds to a person's needs? Well, we're not in a galaxy far, far away. This is happening right here, right now and it's transforming the way we think about education.

PROMOTING SOCIAL INTEGRATION

Remember those school days when the bell rang, signaling recess? The playground buzzed with excitement, friendships blossomed and memories were made. Here's a thought for you—was that experience the same for everyone? Social interactions can be a tad more challenging for some special needs students.

I've always believed that every student, regardless of their abilities, should have the chance to build meaningful relationships with their peers and AI is making this belief a reality. How, you ask?

First up, we've got AI-driven communication tools. Imagine a student with speech difficulties who wants to join a conversation. In the past, they might have hesitated but with these tools, they can express themselves using alternative communication methods, symbols, gestures or even text-to-speech. It's like giving them a voice they never knew they had. And the best part? Their peers can engage right back, breaking down barriers and building bridges.

But AI isn't just about tools and gadgets, it is also about understanding. AI can analyze classroom dynamics, pinpointing moments when a special needs student might feel left out. Then, it suggests activities or groupings that promote inclusivity. Think of it as a behind-the-scenes director orchestrating those heartwarming moments of connection.

This new AI world is one where every student, regardless of their abilities, feels seen, heard and valued. Where classrooms aren't just about lessons and grades but about connections and growth. Sounds dreamy, but with AI, this dream is inching closer to reality.

THE AI ROLLER COASTER

Do you know how we've been raving about the wonders of AI in creating inclusive classrooms? Like any roller coaster, there are thrilling highs and some unexpected drops. So, let's explore the challenges of AI in inclusivity and, more importantly, how we can tackle them head-on.

1. THE CHALLENGE OF BIAS:

First up on our ride is the sneaky issue of bias. AI, bless its digital heart, learns from data. If that data has biases (and let's face it, sometimes it does), AI might unintentionally perpetuate them. Imagine an AI tool that doesn't recognize certain accents or dialects. That's not very inclusive now, is it?

> **The Solution:** We need diverse data sets and continuous monitoring. By feeding our AI tools with varied and representative data, we can ensure they're truly inclusive. Regular check-ins and updates are a must!

2. OVER-RELIANCE ON TECHNOLOGY:

There's a risk that educators might lean too heavily on AI, thinking it's the magic wand for all challenges but always remember, AI is a tool, not a teacher.

> **The Solution:** Balance is key. While AI can offer fantastic support, the human touch that educators bring to the table are Irreplaceable. Let's use AI to enhance, not replace, the human connection.

3. ACCESSIBILITY AND AFFORDABILITY:

Not every school can afford the latest AI tools but even if they can, not all tools are accessible to every student.

> **The Solution:** Collaboration is the name of the game. Tech developers, policymakers and educators need to join forces.

By working together, we can develop affordable, accessible and effective AI solutions for all.

4. PRIVACY CONCERNS:

Last but not least is the loop-de-loop of privacy. With AI collecting and analyzing data, there's always the worry: "Is our students' information safe?"

The Solution: Transparency and strict data protection protocols. Let's be clear about what data we're collecting, why we're collecting it and how we're using it. All while ensuring top-notch security measures.

KEY CHAPTER TAKEAWAYS

Personalized Learning Environments: AI's potential to create classrooms that adjust in real-time to cater to each student's unique needs is groundbreaking.

Promoting Social Integration: Beyond academic support, AI can play a pivotal role in fostering meaningful social interactions for students with special needs.

Challenges and Solutions: While AI offers numerous benefits, it's essential to be aware of potential pitfalls.

CHAPTER LEARNING ACTIVITIES

ACTIVITY 1: AI TOOL EXPLORATION

Research and explore one AI tool designed for special education. Test it out if possible and reflect on its strengths and areas for improvement. How would you integrate it into a classroom setting?

ACTIVITY 2: CLASSROOM DESIGN CHALLENGE

Imagine you're designing a classroom of the future that leverages AI for inclusivity. Sketch or describe the classroom, highlighting at least three AI-driven features that cater to students with diverse needs.

ACTIVITY 3: ETHICAL DEBATE

Organize a discussion or debate on the ethical implications of using AI in special education. Consider data privacy, potential biases and the balance between humans and machines in learning.

8

AI IN HIGHER EDUCATION

Remember the days when we'd sit in massive lecture halls, squinting to see the tiny writing on the chalkboard or the countless hours spent in the library, thumbing through stacks of books for that one elusive reference? Have you ever had a burning question at 2am but no one to ask?

Enter AI-powered chatbots. They are available round the clock, ready to answer queries, schedule appointments, even suggest customized study materials, courses and career paths based on student's digital footprints.

AI works at all levels of education but in spaces of higher education, where university populations can be daunting and administrative tasks are massive, AI really shines—streamlining processes, handling admissions and even predicting which courses will be in demand next semester. The result is smoother operations and more time for educators to educate.

RESEARCH AND AI

I remember my early days in research. I pored over heaps of data, trying to spot patterns, feeling like I was searching for a needle in a haystack. Such is the story of many university students but here's where AI steps in.

Imagine feeding your raw data into a system and the outcome is a beautifully visualized pattern or trend. AI algorithms can sift through vast datasets in the blink of an eye, highlighting correlations and anomalies. It's like having a super-powered magnifying glass that zooms in and points out the cool stuff without all the frustration.

Remember when you wished you could predict future trends based on current data? AI's got your back. With machine learning, these systems can forecast future patterns, giving researchers a head start. It's not quite a crystal ball, but it's the next best thing!

AI isn't just about numbers however. It can also connect researchers working on similar topics and foster collaboration. Think of it as a super-smart academic matchmaker, introducing you to your research soulmates.

Let's face it, research isn't all groundbreaking discoveries. There's a lot of mundane, repetitive work involved. AI can handle literature reviews, data entry and citation management tasks with ease. This means more time for you to do what you love—diving deep into your research.

STUDENT SUPPORT SYSTEMS

Remember your first days on campus, when you felt lost in the maze of a new university? Whether it was choosing the right course, battling those sneaky stress monsters or finding the darn library, we've all been there. What if I told you there's a new buddy on campus, ready to guide, support and even lend a virtual shoulder? Enter AI-driven student support systems.

Lost on campus? Need to know the cafeteria's menu? Or are you just looking for a quiet study spot? AI-driven apps are now your personal campus guides. With real-time data and predictive analytics, they can tell you the best times to hit the gym or the least crowded study spots during finals.

Now imagine, it's enrollment season, and you're torn between "Advanced Quantum Physics" and "History of Renaissance Art" (I know, quite the mix!). Instead of the age-old eeny, meeny, miny, moe, you now have an AI tool that analyzes your academic strengths, interests and career goals to offer personalized course recommendations. It's like having a wise old professor in your pocket, minus the tweed jacket.

What if you need a little help navigating the highs and lows of university life? AI-driven chatbots, equipped with cognitive behavioral therapy techniques, are now available 24/7 to help. They offer a judgment-free zone to vent, seek advice or even chat. Of course, they're not a replacement for professional therapy or real human interaction, but are definitely a step towards breaking the stigma and seeking help.

COLLEGE ADMISSIONS AND ADVISING

THE ADMISSIONS PROCESS

One upon a time, admissions officers had to sift through every college application manually. Thankfully, AI has stepped in to give these hardworking folks a break. With the ability to quickly scan and categorize applications based on set criteria, AI ensures that no detail, however minute, is overlooked. Every aspiring student gets a fair shot, and no application gets lost in the shuffle.

While grades and test scores are crucial, we all know there's more to a student than just numbers. AI tools, with their advanced algorithms, can analyze essays to gauge a student's passion, determination and overall fit for the institution. It's like having extra eyes to see the heart and soul behind the words.

Here's where it gets super interesting though. Based on past data, AI can predict how likely a student will be to succeed at a particular institution as well. This isn't about weeding people out but ensuring that students find the right fit where they'll thrive.

With predictive analytics, educators can get insights into which students might need extra help or resources. It's like having a heads-up to intervene early and make a difference.

It's not just about academic scores. These tools consider various factors, from attendance records to participation in extracurricular activities. It's a holistic approach to understanding students' journeys.

On the flip side, identifying potential dropouts early means we can provide targeted support to those students. Maybe they need counseling, financial aid or academic tutoring. By pinpointing the challenges, we can address their needs head-on.

Now, I won't lie. Predicting a student's academic future might sound a tad bit sci-fi. But trust me, it's all grounded in data and genuine concern for each student's success.

AI-POWERED CAREER GUIDANCE

Have you ever had that moment when you're staring at a course catalog, feeling overwhelmed? Or perhaps you've found yourself wondering, "Is there a job out there that matches my passion for underwater basket weaving?" (Okay, maybe not that specific, but you get the gist.)

Remember those old dating shows where contestants would answer questions, and a computer would find their "perfect match"? Well, imagine that, but for your career. These AI tools analyze your strengths, interests and even quirks to suggest courses that align with potential career paths. It's personalized advice without unsolicited opinions!

The job market is a fickle beast, constantly evolving, with new roles popping up as others fade. AI doesn't just look at what's hot now but forecasts market demand. So, if there's a rising demand for virtual reality developers or sustainable agriculture experts, you'll be the first to know!

These tools provide guidance, but there's always room for exploration and change. Life's a journey, and sometimes the detours are the best parts.

LIFELONG LEARNING WITH AI

LIFELONG EDUCATION AND SKILLS DEVELOPMENT WITH AI

When we were kids, every day was a new adventure. We'd learn something new, whether riding a bike or tying shoelaces. Fast forward to adulthood, and the learning never really stops.

Times change, and with it, the skills we need. Today, it's all about digital literacy, data analysis and understanding AI but how do we keep up? Enter AI-driven learning platforms based on your interests, strengths and identified areas of improvement, AI crafts a learning path just for you. Personalized, efficient and always up-to-date.

The beauty of AI is that it's not confined to a classroom. With smartphones and laptops, learning can happen anywhere, anytime. Late-night coding session? Check. Lunch break language lesson? Why not! AI adapts to your pace, making learning flexible and fun.

Here's the thing about the future—it's unpredictable. Thankfully with AI, we can get a sneak peek. By analyzing job market trends, AI can predict the skills in demand. This way, you are prepared for today and also ready to tackle tomorrow's challenges.

UPSKILLING AND RESKILLING

Imagine you're a sailor, and you've mastered sailing in calm waters but suddenly, the winds change, and you face turbulent seas. What do you do? You adapt, learn new techniques and sail through the storm. Now, replace the seas with the job market and the changing winds with technological advancements.

That's where we are. Just like the sailor, we need to upskill and reskill to navigate these waters.

Many are scratching their heads and wondering why the job market is changing. It seemed like yesterday when typewriters were all the rage but students of today have no concept of what a typewriter even looked like. Gone are the days when being proficient in Microsoft Word was a big deal on your resume. With AI, automation and other tech advancements, some of the job skills of yesterday are evolving, while others have become obsolete.

Think of upskilling as adding more tools to your toolkit. It's about building on what you already know. Say you're a graphic designer familiar with basic software. Upskilling might mean mastering advanced design tools or even dabbling in animation. It's like going from making delicious spaghetti to whipping up a full Italian feast!

Reskilling is like switching from playing basketball to soccer. It's about learning a completely new set of skills. Maybe you've been in retail for years, but you've always had a passion for digital marketing. Reskilling would be diving headfirst into that world, learning the ropes and starting a new chapter in your career.

Change can be daunting, I get it, but it's also exciting. It's a chance to grow, to challenge ourselves and to dive into fields we'd never imagined. With resources, online courses and AI-driven platforms, the world is our oyster.

KEY CHAPTER TAKEAWAYS

Personalized Learning Paths: AI offers tailored educational experiences, understanding that every student's learning journey is unique.

Research Enhancement: AI aids in data analysis and predictive trends and fosters collaboration, making the research process more efficient and insightful.

Student Support Systems: From academic advising to mental health support, AI provides round-the-clock assistance, ensuring students are well-supported throughout their academic journey.

Lifelong Learning: AI promotes continuous learning, helping individuals upskill and reskill as the job market evolves.

CHAPTER LEARNING ACTIVITIES

ACTIVITY 1: AI EXPLORATION

Explore an AI-driven educational platform or tool. Spend a week using it and journal about your experiences. What did you like? What felt challenging? How did it enhance your learning?

ACTIVITY 2: DEBATE

Organize a group discussion or debate on the ethical implications of AI in higher education. Consider data privacy, potential biases and the balance between technology and human touch.

ACTIVITY 3: CAREER FORECASTING

Use an AI-driven career guidance tool to analyze your current skills and interests. Based on its recommendations, research one career path in-depth. What skills are required? How is the industry evolving? What role does AI play in that field?

9
AI AND STUDENT ENGAGEMENT

Ever wondered why games are so addictive? How can we harness that power to make education more engaging? The thrill of the chase, the joy of discovery and the sweet taste of victory (or, occasionally, the playful lament of defeat). That's the essence of gamification—tapping into our innate love for play and challenges. But why do we love games so much?

It's all about the brain. When we play, our brains release dopamine, that feel-good neurotransmitter. The same chemical gets released when we achieve a goal, solve a puzzle or overcome a challenge. Learning is all about challenges and achievements. By gamifying the learning process, we're essentially giving our brains a dopamine double-whammy. We are not just learning, we're enjoying the journey and when we enjoy something, we're more likely to stick with it and succeed.

Imagine playing a game that adapts to your strengths and weaknesses, offering you challenges that are just right—not too easy or too hard. That is what AI brings to the table.

For instance, think about a language learning app. Instead of generic quizzes, AI can analyze your progress and tailor questions in the areas you struggle with. Past tense verbs got you confused? The AI just might throw in more challenges around those to bring you up to speed. Aced the vocabulary section? It'll give you advanced words to keep you on your toes.

AI can also introduce dynamic storytelling. Think of a history lesson transformed into an epic adventure game. As you make choices, the story evolves and you learn about historical events based on your decisions. It's immersive, interactive and incredibly engaging.

Have you ever played those video games with super realistic graphics and thought, "Wow, this feels real"? That's AI at work, analyzing vast amounts of data to render lifelike images. Now, picture that level of realism in educational games. We're talking about virtual field trips to ancient civilizations or exploring the human body from the inside. The possibilities are endless!

SUCCESS STORIES

1. Language Quest—This platform took the world by storm with its immersive language-learning adventures. The AI tailors the difficulty based on your progress. So, as you master the basics, the challenges get trickier, pushing you to learn more. Thousands of users went from "Bonjour" to discussing French literature in months.

2. Eco Explorer—This platform transports students to virtual ecosystems, from dense rainforests to arctic tundras. As they navigate these worlds, they encounter challenges related to environmental science. The AI observes their choices, offering feedback and adapting scenarios. It's like having a personal ecology tutor in a video game format!

GAMIFIED LEARNING ASSESSMENT

Assessing learning within gamified environments is an innovative approach that combines education and entertainment, often called "edutainment."

Gamified environments engage students by incorporating game design elements in learning contexts. This approach can motivate students by offering rewards, competition and a sense of achievement, which are not always present in traditional learning environments.

Unlike traditional assessments that occur periodically, gamified learning allows for continuous, real-time assessment. Students' actions can be measured and evaluated by interacting with the game. This provides immediate feedback to the student and valuable data for educators.

In a gamified learning environment, AI can track various metrics such as decision-making processes, response times and the application of knowledge. This data can be used to provide personalized feedback to students, highlighting their competencies and identifying areas where they need more practice.

Based on the assessment data, gamified platforms can adapt to each student's learning needs. If a student excels in certain areas, the game can present more challenging scenarios. Conversely, if a student struggles, the game can provide additional practice or simplify concepts to match the student's pace.

Some advanced gamified platforms incorporate AI tools that analyze facial expressions and body language to gauge a student's emotional state. This can be particularly useful in understanding the student's level of engagement, frustration or confusion, allowing the game to respond appropriately, much like an attentive teacher would.

Gamified environments often place students in immersive scenarios that require applying knowledge in context. For example, a student studying medicine, may participate in a role play where they act as a doctor diagnosing patients. This allows the student to apply medical knowledge in a simulated real-world setting, making the learning experience more relevant, engaging and safe for all involved. These environments also allow for assessing soft skills such as problem-solving, critical thinking and teamwork. As students navigate the game, they can develop and refine these skills, which are crucial for success in the real world.

While gamified learning environments offer many benefits, there are challenges. Ensuring that the game accurately assesses learning outcomes, aligns with educational standards and remains accessible to all students is important. As with any AI-driven system, ethical and privacy concerns must also be addressed. As mentioned, it is essential to ensure that all player data collected is secure, used responsibly and with the consent of the students and their guardians.

AI AND SOCIAL LEARNING

Remember those group projects in school? The ones where you'd huddle with classmates, brainstorming ideas and occasionally get sidetracked by the latest school gossip? Ah, good times! Those group projects—the debates and the discussions—weren't just fun distractions. They were pivotal to our learning experience.

You see, learning is not just about absorbing facts, it's a social endeavor. We humans are wired to learn from one another. It's in our DNA. From the ancient tribal gatherings around a fire to modern-day classroom discussions, we've always learned best when we collaborate and interact. Why?

When we share ideas, challenge each other's viewpoints and build on one another's knowledge, we gain a deeper understanding of the subject. It's like piecing together a jigsaw puzzle with friends—each person adds a piece and as they do, the image becomes clearer.

Now, I know what you're thinking. "That's all well and good, but what about the digital age? With online classes and e-learning, aren't we missing out on the social aspect?" Well, that's where AI comes into play. Enter the world of AI-driven collaborative platforms. These aren't your average chat rooms or forums. They're much more sophisticated than that.

Imagine a virtual classroom where students from around the world come together. They're working on a project, where the AI tool is a facilitator. It suggests resources, connects students with similar interests and even identifies gaps in the group's knowledge, prompting further research. Some platforms even use AI to analyze group dynamics. If one student dominates the conversation, the AI might prompt quieter members to share their thoughts. It might even suggest resources to help resolve the debate if there is a group disagreement.

Remember those escape rooms that were all the rage a few years ago? Think of these AI tools as a virtual escape room. You're collaborating and solving problems, as each person brings something unique to the table, and just like in an escape room, the thrill of solving a challenge together is unmatched.

BUILDING VIRTUAL LEARNING COMMUNITIES

Have you ever been part of an online community? Maybe it was a fan forum, a hobbyist group or a digital book club? There's a certain beauty in connecting with strangers over shared interests, forming bonds, without ever meeting face-to-face. Now, translate that to an educational setting and that's what we are talking about when we say "virtual learning communities."

Now, I've seen some folks raise an eyebrow at this. "How can an online community replace the real-life classroom experience?" they ask. Well, with the help of AI, it's not just possible; it's revolutionary. AI can analyze students' learning patterns, interests and strengths in order to group them into communities where they'll thrive. Imagine a digital space where a student from Tokyo collaborates on a project with peers from Nairobi, London and Rio. The diversity of thought, the exchange of ideas—it's a melting pot of knowledge!

AI doesn't just stop at forming these communities but it also nurtures them. It can suggest discussion topics, prompt debates and even detect when a member feels left out, ensuring everyone's voice is heard. It is like having a community manager who never sleeps!

CHALLENGES AND SOLUTIONS

Now, I won't sugarcoat it. Like any great adventure, there are challenges. How do we ensure these AI-driven platforms are inclusive and don't perpetuate biases? And what about the digital divide? Let's face it, not everyone has access to high-speed internet or the latest gadgets but with every challenge comes an opportunity. For biases, we can train AI with diverse datasets, ensuring it understands and respects human experiences. As for the digital divide, partnerships between tech companies, governments and educational institutions can bridge the gap, providing resources to those in need.

And let's not forget the human touch. While AI can facilitate, educators, students and parents breathe life into these communities. It's their passion, their commitment, and their drive that makes the magic happen.

ANALYZING STUDENT BEHAVIOR

Have you ever watched one of those detective shows where they piece together clues to solve a mystery? In education, we have our version of that and it's all about understanding our students better. Let's talk about data, but not the boring spreadsheet kind. I'm talking about the stories, the patterns and the little "aha!" moments that give us insights into how students learn.

What if every time a student interacts with an online learning platform, they leave behind a digital footprints? Such footprints would tell educators the amount of time they spend on a particular topic, the number of attempts they make on a question or even the patterns of their breaks. If I ask you, "What can you learn from this?" you might shrug but here's where something amazing happens. By collecting and analyzing this data, we can get a deeper understanding of each student's learning journey. It's like a magnifying glass that lets us see their struggles, victories and everything in between. This is not about collecting data willy-nilly. It's about doing it in a way that genuinely benefits each learner individually.

PREDICTIVE ANALYTICS

Imagine a student, let's call her Mia. Mia's been struggling with a particular topic and her engagement levels have dropped. Now, in a traditional setting, Mia might slip through the cracks but with predictive analytics, alarms would be ringing. The system would identify Mia as an "at-risk" student, not because she's incapable, but because she needs a different approach or additional support.

Here's where it gets even cooler, however. The system doesn't just raise the alarm on Mia; it suggests interventions. Maybe Mia needs more visual content or perhaps a one-on-one session with a tutor would help. The goal is to ensure that every Mia out there gets the support they need, when they need it.

But, and this is crucial, it's not about labeling or pigeonholing students. It's about empowerment. It's about giving educators the tools to make a difference. At the heart of all of this stands a simple belief—every student deserves a chance to shine.

ETHICAL CONSIDERATIONS

With great power comes great responsibility, which couldn't be truer in AI and education. While we're all excited about the potential of AI, there's a crucial aspect I must continue to emphasize: ethics.

First and foremost, student privacy. Remember those digital footprints we talked about earlier? While they offer invaluable insights, we are also responsible for protecting student data. We're talking about young minds here and their trust is paramount.

Then there's the issue of responsible data use. It's easy to get lost in the sea of data and forget that behind every data point is a real person with emotions, aspirations and dreams. We must ensure that AI tools are used to uplift, not pigeonhole or stereotype.

KEY CHAPTER TAKEAWAYS

Gamification and AI: The combination of gamification and AI offers a personalized, adaptive learning experience, making education more engaging and effective.

AI-Powered Collaborative Tools: AI fosters global collaboration, creating virtual learning communities where students from different backgrounds can share and learn from each other.

Analyzing Student Behavior with AI: By tracking digital footprints, educators can gain insights into individual learning journeys, allowing for timely interventions and support.

Ethical Considerations: The use of AI in education comes with ethical responsibilities, especially concerning student data privacy and responsible data use.

CHAPTER LEARNING ACTIVITIES

ACTIVITY 1: DESIGN YOUR OWN AI-DRIVEN GAME

Think of a topic you are passionate about and design a basic game that uses AI to adapt to the player's knowledge level. Sketch out the game's structure, challenges and how AI would adjust the game play.

ACTIVITY 2: JOIN OR CREATE A VIRTUAL LEARNING COMMUNITY

Explore online collaborative learning platforms. Join a group or start your own, focusing on a topic of interest. Reflect on how AI tools could enhance the collaborative experience.

ACTIVITY 3: ETHICAL DEBATE

Engage in a group discussion or debate the ethical implications of using AI in education. Consider data privacy, potential biases in AI algorithms and the balance between personalization and pigeonholing.

10

THE GLOBAL IMPACT OF AI IN EDUCATION

I often find myself daydreaming about a world where every child, no matter where they are born, has equal access to quality education. A world where the zip code of your birth doesn't dictate the quality of your schooling. Sounds like a fairy tale, right? With AI, this dream is inching closer to reality.

Imagine a remote village between mountains, where the nearest school is miles away. Now, picture a child from this village, with a sparkle in their eyes, holding a tablet. This isn't just any tablet, though, it is one powered by AI. The tablet is a window to the great big world out there and tailored to the child's learning style and full of interactive lessons. The child, who once had limited access to textbooks, now has a world of knowledge at their fingertips.

This is not just about access however, it's about quality. In many developing regions, there's a shortage of trained teachers. AI can step in here and cover the shortfall—24 hours a day, 7 days a week—ensuring no child is left behind.

CHALLENGES IN IMPLEMENTATION

As much as I would love to paint a rosy image of AI seamlessly integrating into every classroom globally, we've got some hurdles to jump over first. Trust me, they are not just tiny speed bumps, but full-blown roadblocks!

Infrastructure: Imagine running the latest video game on a computer from the 90's. Sounds frustrating, right? That's the challenge many schools in developing regions face. They might not have stable electricity, high-speed internet or modern devices. Without these basics, how can we even begin to think about introducing AI?

Funding: It's an age-old challenge. Implementing AI-driven solutions isn't cheap. We're discussing investing in hardware, software, training and maintenance. This is a tall order for many countries already grappling with budget constraints and while we're on the topic, who will foot the bill? Governments? Private sectors? NGOs? It's a puzzle that needs solving.

Cultural barriers: Every region and every community has its own set of beliefs and values. Introducing new technology can sometimes clash with this. There might be resistance, skepticism or even fear. It is crucial to approach this with sensitivity, understanding and a whole lot of patience.

Before you get disheartened, remember this: every big change comes with challenges. While the road ahead might be bumpy, the destination—a world where every child has access to quality education—is worth every twist and turn.

SUCCESS STORIES

Amidst all the challenges and roadblocks, this shining hope makes you go, "Wow, we're onto something here!" Let me share a couple of heartwarming tales. First, let's talk about rural India. Visualize a tiny classroom with eager kids but just one teacher for all subjects. Enter an AI-powered app that personalizes learning for each student. The result is that kids who were once struggling now outperform their urban counterparts and the best part is—the app works offline, considering the spotty internet connection. It's like having a personal tutor in your pocket.

Now, let's hop over to Africa. In parts of Kenya, schools are few and far between. But thanks to a nifty AI-driven program, students can access lessons via mobile phones, and we're not just talking about reading and writing. They're learning to code! From the heart of the Savannah, these kids are gearing up to be the tech leaders of tomorrow.

In South America, a village in Brazil uses AI to translate educational content into local indigenous languages. This ensures that learning isn't limited to those who speak Portuguese but reaches everyone, preserving culture while promoting education.

These stories are just the tip of the iceberg. They show that with a dash of innovation, a sprinkle of determination and a whole lot of heart, AI can truly transform lives.

FUTURE PROSPECTS

I've got my crystal ball here (well, not really, but let's pretend), and I'm seeing some pretty exciting stuff on the horizon for

AI in education, especially in developing countries. Trucks equipped with all the tech goodies drive into the remotest villages and bring education to doorsteps. These trucks won't have computers but AI systems that can adapt to each child's learning style. Think of them as pop-up schools but on steroids.

There are thousands of dialects worldwide, many in developing regions. I foresee AI-driven apps that can instantly translate educational content into any dialect, ensuring that every child, no matter where they're from, gets to learn in their mother tongue.

Not to replace our beloved human educators, of course, but to assist them. These AI entities could be available 24/7, answering questions, providing resources and ensuring that learning never stops, not even for a second.

Lastly, I see a massive surge in collaborative projects. Schools worldwide connect, share resources and work on projects together, all facilitated by AI. It's like a global classroom without borders.

CULTURAL CONSIDERATIONS

As we explore an AI-driven world, the beautiful cultures, traditions and histories of our global community often get overlooked. Trust me, when it comes to implementing AI in education, this is a biggie!

Imagine you're introducing a fantastic new AI tool in a classroom. It's sleek, it's efficient and it's cutting-edge but the content it delivers is entirely out of sync with the local culture.

Instead of engaging students, it alienates them. It's like trying to fit a square peg into a round hole. Not the best scenario, right?

Remember when you traveled to a new place and tried to use a phrasebook to communicate? Sure, you got the words right, but sometimes the context was all wrong, leading to pretty hilarious (or embarrassing) situations. That's precisely what happens when we don't adapt AI tools to fit the local context.

Every culture and tradition has nuances, stories and values, if we want AI to truly revolutionize education globally, we must ensure it respects and understands these nuances. It's not just about translating content into different languages, it is about adapting the content to resonate with the hearts and minds of the students.

LANGUAGE AND LOCALIZATION

Picture yourself in a market in Marrakech. The air is filled with the aroma of spices and you're trying to buy a beautiful handmade rug—but you only speak English and the vendor only speaks Arabic. You both have a fantastic tool that translates words, but it doesn't quite capture the essence or the emotions of what you are trying to say. Frustrating, right?

Now, let's pivot back to our world of education. We've got these shiny AI tools promising to revolutionize learning but are they serving everyone if they' are only available in one language or don't consider the cultural nuances of your local? I'd argue, not quite.

Language is more than just words; it's a reflection of our identity, culture and history. When we talk about localizing AI tools, it's not just about translating content. It's about making sure the content feels right and homey. It's about ensuring that a student in Tokyo gets the same 'aha!' moment as a student in Buenos Aires.

ADDRESSING BIAS

I once tried cooking a dish from an online recipe and let's say it didn't turn out as expected. Why? Because I made assumptions based on my own experiences and didn't consider the cultural context of the dish.

Similarly, when we talk about AI in education, we're dealing with a recipe written by humans and, just like me with my culinary misadventure, we can inadvertently introduce our biases into these tools if we're not careful. It's like adding too much salt to a dish—it can overshadow the genuine flavors.

CULTURAL CASE STUDIES

1. The African Math Whiz App: It's an AI-powered app developed specifically for students in rural African communities. The creators didn't just slap on some algorithms and call it a day. They went deep, incorporating local folklore and stories into math problems. Imagine solving a math equation based on a local legend! Not only did it make learning fun, but it also resonated with the student's cultural background. Talk about hitting two birds with one stone!

2. The Language Bridge in Asia: In several Asian countries with multiple dialects, an AI tool was designed to help students learn the national language but it used their local dialects as a bridge. So, instead of feeling alienated, students felt seen and heard, and their learning curve skyrocketed!

3. Indigenous Wisdom Meets Modern Tech in South America: In the Amazon, an AI tool was crafted to help indigenous communities document and share their ancestral knowledge. The tool was sensitive to their oral traditions, allowing elders to "speak" to the system, translating their wisdom into interactive lessons for the younger generation.

Isn't it fascinating how it becomes so much more powerful when technology is molded with a touch of cultural essence? These case studies testify that AI in education can truly shine when we respect and integrate local traditions and values.

COLLABORATIVE INTERNATIONAL EFFORTS

You know that feeling when you are trying to solve a puzzle and you're stuck on one piece? Then, someone walks by, takes a glance, and says, "Hey, try flipping that piece around..." and voila! It fits. That's the magic of collaboration. Now, imagine that globally, with countries sharing their knowledge and resources. Mind-blowing, right?

When countries come together, they're not just pooling resources; they're pooling experiences, insights and innovations. Think of it as a potluck dinner where everyone brings a dish. Some countries might bring cutting-edge technology, while others bring invaluable local insights. The result is a feast of knowledge that's richer and more diverse than what any single country could whip up alone and here's the cherry on top: shared challenges, like limited resources or infrastructure, can be tackled more efficiently. Instead of reinventing the wheel, countries can adapt solutions that have worked elsewhere. It's like getting a cheat sheet for the most complex exam ever!

KNOWLEDGE EXCHANGE

Knowledge exchange is about sharing the good stuff: best practices, groundbreaking research and those "Aha!" innovations. Why? Because when we share, we grow. Think about it—if one country discovers a fantastic way to use AI to boost reading skills, why should they keep it to themselves? By sharing, they can help students worldwide.

And here's another thing. When we exchange knowledge, we also learn from others' mistakes. It's like getting a heads-up about which dishes at the potluck might be too salty. This way, we can avoid pitfalls and fast-track our way to success.

FUTURE PROSPECTS

What does the future hold for international collaborations in AI-driven education? Well, for starters, I see borders becoming blurrier. With technology bridging gaps, countries will collaborate more seamlessly, like neighbors swapping gardening tips over the fence.

Next, I predict a rise in global AI education conferences, workshops and maybe even AI-themed festivals! These events will be melting pots of ideas where educators, policymakers and techies come together to dream big.

Lastly, as AI tools become more advanced, I foresee a global digital library accessible to all, brimming with AI-driven educational resources. It'll be like having a universal school, where everyone's invited.

KEY CHAPTER TAKEAWAYS

Equal Access to Education: AI has the potential to bridge educational gaps, ensuring every child, regardless of their location or circumstances, has access to quality education.

Challenges and Solutions: While the promise of AI in education is vast, challenges like infrastructure, funding and cultural barriers need to be addressed. However, with collaboration and innovation, these challenges can be overcome.

Cultural Considerations: AI tools must be sensitive to local cultures and traditions to be effective. It's not just about translating content; it's about making the content resonate with the learners.

CHAPTER LEARNING ACTIVITIES

ACTIVITY 1: RESEARCH AND REPORT

Choose a developing country and research the state of its education system. How could AI be integrated to improve educational outcomes? Prepare a short report or presentation on your findings.

ACTIVITY 2: DESIGN A LESSON

Imagine you're creating an AI-powered lesson for students in a specific region. Design a lesson that incorporates local folklore, traditions or stories. Ensure the lesson is interactive and engaging.

ACTIVITY 3: DISCUSSION FORUM

Join an online forum or group focused on AI in education. Share your insights from this chapter and engage in discussions. What new perspectives or ideas did you gain from interacting with others?

PART 2

EMERGING TECHNOLOGIES AND THE HORIZON OF EDUCATION

11

EMBRACING AI IN MODERN EDUCATION

The education sector, traditionally seen as a bastion of human interaction and personal touch, has not remained untouched by the AI wave. Schools, colleges and universities increasingly leverage AI to enhance administrative functions and the learning experience. The potential of AI in education is vast, from personalized learning paths based on individual student's needs to automating administrative tasks, thereby allowing human educators to focus more on teaching and less on paperwork.

In a move that captured headlines and sparked worldwide debates, a top private school appointed an AI chatbot as its headteacher. This wasn't just a symbolic gesture or a marketing gimmick. It was a testament to the school's faith in the capabilities of AI and a nod to the future of educational leadership. This bold move marked a significant milestone in the journey of AI in education, highlighting the potential of AI not just as a tool for teaching but as a leader and decision-maker.

The appointment of an AI chatbot as the 'headteacher' at a prestigious private school is not an isolated incident but a reflection of a larger trend. As this chapter progresses, we will explore various such initiatives where AI plays a pivotal role in reshaping the education landscape. From administrative tasks to classroom teaching, from decision-making to student engagement, AI's influence is pervasive and transformative.

THE CHANGING FACE OF EDUCATIONAL LEADERSHIP

Traditionally, education administration and leadership have been dominated by human intuition, experience and judgment. From setting the vision and mission of an educational institution to making critical decisions that impact students and staff, human leaders have always been at the helm.

However, with the advent of AI, this dynamic is transforming. With its data-driven insights and predictive capabilities, AI is gradually complementing human judgment with machine precision.

CASE STUDY: COTTESMORE SCHOOL Nestled in the picturesque county of West Sussex, Cottesmore School has long been known for its commitment to excellence in education. With a history of over a century, the school has always prided itself on its ability to evolve with the times, embracing innovation while staying true to its core values. It's this forward-thinking approach that led Cottesmore to explore the potential of AI in education leadership.[1]

THE APPOINTMENT OF ABIGAIL BAILEY, AI 'PRINCIPAL HEAD TEACHER'

In an audacious and visionary move, Cottesmore School introduced Abigail Bailey, an AI chatbot, as its 'principal headteacher'. While the idea of an AI head teacher might seem futuristic to many, it was a logical step in Cottesmore's innovation journey. Abigail Bailey wasn't just a fancy addition to the school's tech arsenal; she was a strategic decision to enhance the school's administrative and leadership capabilities.[2]

HOW ABIGAIL BAILEY AIDS THE HUMAN HEADMASTER

Abigail Bailey's introduction to the school was not to replace the human headmaster but to augment his capabilities. With her advanced algorithms and vast data processing abilities, Abigail assists in various administrative tasks, such as:

Data Analysis: Analyzing student performance metrics, attendance records and feedback to provide insights that can shape teaching methodologies and curricula.

Resource Allocation: Optimizing the allocation of resources, from classroom assignments to staff scheduling, ensuring maximum efficiency.

Communication: In drafting communications to parents, staff and other stakeholders, ensuring clarity and consistency.

Decision Support: Providing data-driven recommendations on critical decisions, from student admissions to staff hiring.

BENEFITS OF AI IN ADMINISTRATION

From supporting staff to drafting school policies, the integration of AI in education administration offers a plethora of benefits. These include:

Efficiency: Automation of routine tasks reduces administrative overhead, allowing human staff to focus on more strategic initiatives.

Precision: AI's data-driven approach ensures that decisions are based on concrete evidence rather than mere intuition.

Proactivity: Predictive analytics enables the school to anticipate and address challenges before they escalate.

Inclusivity: AI can help craft policies that cater to the diverse needs of the student population, ensuring that no student is left behind.

Continuous Learning: Just as students learn and evolve, AI systems like Abigail Bailey continuously learn from data, becoming more refined and effective.

Integrating AI in leadership roles, especially in sectors as personal and human-centric as education, often raises eyebrows. The common misconception is that AI might overshadow or replace human judgment, intuition and emotion. However, the true essence of AI's role in leadership is not replacement but collaboration. It's about creating a harmonious blend where AI amplifies human capabilities, and humans provide the emotional and ethical compass that machines lack.

AI COLLABORATION: ROGERSON'S PERSPECTIVE

Tom Rogerson, the human headmaster of Cottesmore School, has been at the forefront of this AI revolution in education leadership. His perspective on collaborating with AI, particularly with Abigail Bailey, is enlightening and pragmatic.

For Rogerson, AI is a tool, a valuable assistant that provides data-driven insights, automates mundane tasks and offers recommendations. However, the final decisions, especially those with deep emotional or ethical implications, always rest with him. He views Abigail Bailey not as a competitor but as a collaborator, a digital colleague who complements his skills and fills the gaps.

ADDRESSING THE LONELINESS OF LEADERSHIP

Leadership, especially in pivotal roles like that of a headmaster, can often be lonely. The weight of responsibility, the constant need to make critical decisions and the emotional toll of managing a large institution can be overwhelming. While colleagues, friends and family can offer support, there are aspects of leadership that only someone in that position can truly understand.

Surprisingly, this is where AI can play a role. While AI lacks emotions, it can provide a form of support that's unbiased, non-judgmental and purely based on data. For Rogerson, having Abigail Bailey as a sounding board, a digital entity that can provide insights without any emotional baggage, has been invaluable. It's not about replacing human emotional support but adding analytical support.

Furthermore, AI can help identify patterns that might indicate stress, burnout or other challenges among staff members, allowing leaders to take proactive measures. It can serve as an early warning system, ensuring the well-being of the institution's most valuable asset: its people.

Beyond the analytical support, AI can also assist leaders in managing their time, prioritizing tasks and even self-care. By automating administrative tasks, AI can free leaders to focus on self-reflection, professional development and building deeper human connections.

While integrating AI in teaching and administrative roles has garnered significant attention, there's another dimension to AI's involvement in educational institutions: strategic planning and decision-making. As schools and colleges recognize the transformative potential of AI, there's a growing emphasis on creating dedicated roles that focus on harnessing AI's capabilities for long-term institutional growth and innovation.

COTTESMORE'S SEARCH FOR A 'HEAD OF AI'

Having already made headlines with its appointment of an AI 'headteacher', Cottesmore School took another pioneering step in its AI journey. Recognizing the need for a dedicated role overseeing the school's AI initiatives, Cottesmore searched for a 'Head of AI'. This role was envisioned not just as a technical position but as a strategic one, bridging the gap between AI's capabilities and the school's vision for the future.

In a move that further solidified Cottesmore's position as a trailblazer in AI-driven education, the school introduced Jamie Rainer, an AI entity, as its 'Head of AI'. Unlike traditional AI systems designed for specific tasks, Jamie Rainer was conceptualized as a strategic thinker, an AI entity capable of long-term planning, trend analysis and visionary thinking.

Jamie Rainer's appointment was more than just a symbolic gesture; it marked a significant shift in how educational institutions perceived AI's role. Here's a glimpse into Jamie's multifaceted responsibilities:

- **Trend Analysis:** Leveraging vast datasets to identify emerging trends in education technology, ensuring that Cottesmore stays ahead of the curve.
- **Strategic Planning:** Collaborating with human stakeholders to draft a long-term AI strategy for the school, aligning AI initiatives with the institution's broader goals.
- **Resource Allocation:** Recommending investments in AI research, tools and training, ensuring optimal utilization of resources.
- **Ethical Oversight:** Ensuring that all AI initiatives adhere to ethical standards, safeguarding the interests of students, staff and other stakeholders.
- **Collaboration:** Building bridges with AI researchers, institutions and industry experts, fostering a culture of collaboration and continuous learning.

As we stand on the cusp of a new era, it's evident that AI is not just a fleeting trend or a niche technology. It's rapidly becoming an integral part of our daily lives, reshaping industries and redefining societal norms. In such a landscape, the younger generation must not be passive consumers of AI but active collaborators, understanding its nuances and harnessing its potential.

THE IMPORTANCE OF STUDENTS LEARNING TO WORK ALONGSIDE AI

- **Skill Development for the Future:** As the job market evolves, many of the roles that today's students will take on in the future will involve working alongside AI. Students will be better equipped to navigate their professional journeys by understanding AI's capabilities and limitations.

- **Critical Thinking and Ethical Considerations:** AI, while powerful, is not infallible. Students must develop critical thinking skills to question, validate and ethically use AI outputs.

- **Fostering Creativity:** With AI handling repetitive and mundane tasks, students can focus on higher-order thinking, innovation and creativity, exploring areas where human intuition and AI's precision can create magical things.

- **Building a Collaborative Mindset:** The future is not man vs. machine but man and machine. By learning to collaborate with AI, students develop a mindset of synergy, understanding the value of diverse inputs and collaborative problem-solving.

ROGERSON'S VIEW ON AI

With his firsthand experience integrating AI into Cottesmore School's fabric, Tom Rogerson has a unique perspective on AI's role in education and beyond. For Rogerson, AI is not just another tool in the educator's toolkit; it's a 'world-changing' technology.

He believes that AI has the potential to bridge educational gaps, democratize access to quality education and personalize learning experiences like never before. But beyond these tangible benefits, Rogerson emphasizes the transformative impact of AI on students' mindsets. By interacting with AI, students learn the values of adaptability, continuous learning and resilience, preparing them for a dynamic and ever-evolving future.

Rogerson also stresses the importance of ethical considerations. As AI becomes more pervasive, students must understand the ethical implications of AI decisions, ensuring that technology is used responsibly and equitably.

WILL ROBOTS REPLACE HUMAN TEACHERS?

The short answer is no. While AI can automate tasks, analyze vast amounts of data and even simulate human-like interactions, it lacks the intrinsic human qualities that form the bedrock of education. Empathy, intuition, moral judgment and the ability to understand and cater to the emotional needs of students are uniquely human attributes. AI is designed to augment these human qualities, not replace them. For instance:

- **Personalized Learning:** While AI can tailor learning materials based on a student's performance, it's the human teacher who provides the emotional support, encouragement and motivation that students need.
- **Administrative Tasks:** AI can handle administrative tasks like grading and attendance, allowing teachers to spend more time on student interaction, mentorship and personal development.
- **Data-Driven Insights:** AI can provide insights into student performance, but it's the human teacher who devises strategies, interventions and support mechanisms to address individual challenges.

THE BALANCE BETWEEN TRADITIONAL EDUCATION VALUES AND MODERN TECHNOLOGICAL ADVANCEMENTS

Education, at its core, is a deeply human endeavor. It's about imparting knowledge, but it's also about shaping character, instilling values and nurturing the next generation. These fundamental principles of education remain unchanged, even in the face of technological advancements.

However, modern technology, especially AI, offers tools to enhance traditional education values. With AI-powered platforms, students can collaborate with peers from across the globe, fostering a sense of global citizenship and broadening horizons.

AI-driven adaptive learning platforms ensure that learning is not confined to the classroom. Students can learn independently, revisiting concepts and exploring advanced materials as needed. As students interact with AI, they also learn about the ethical implications of technology, preparing them to be responsible digital citizens.

KEY CHAPTER TAKEAWAYS

AI as a Collaborator, Not a Competitor: The essence of AI in education is not about replacing the human touch but augmenting it. From personalized learning to administrative tasks, AI enhances human educators' capabilities.

The Human-AI Synergy: The future of education lies in the harmonious collaboration between human intuition and AI's precision. While AI can provide data-driven insights and automate tasks, education's emotional, ethical and intuitive aspects remain inherently human.

Ethical and Responsible Use of AI: As AI becomes more integrated into the educational landscape, there's a pressing need to ensure that technology serves all students equitably and responsibly.

CHAPTER LEARNING ACTIVITIES

ACTIVITY 1: DEBATE SESSION

Organize a debate on the topic, "AI in Education: Boon or Bane?" This will allow readers to articulate their understanding of the chapter and appreciate diverse viewpoints.

ACTIVITY 2: AI IN YOUR SCHOOL

Imagine you're the principal of a school. Draft a proposal on how you would integrate AI into your institution, considering both its academic and administrative applications. Ensure you address potential challenges and ethical considerations.

ACTIVITY 3: INTERVIEW WITH AN AI

Using any AI chatbot platform available online, engage in a conversation. Ask the AI about its role in education, its limitations and its vision for the future. Reflect on the experience and jot down your observations.

12
OVERCOMING CHALLENGES AND PITFALLS

COMMON CHALLENGES:

RESISTANCE TO CHANGE

Ever tried convincing a kid to eat broccoli for the first time? Or maybe you've attempted to introduce your grandma to the world of smartphones? It's tough. That's because humans, by nature, are creatures of habit. We find comfort in the familiar and anything new can be intimidating. Now, let's talk about introducing AI in education. It's like getting the whole world to try a new dish simultaneously!

First off, there's the big 'F'—Funding. Imagine wanting to bake a fancy cake but realizing you're short on ingredients and cash. Schools and institutions often face this dilemma. They see the potential of AI, but the price tag is a bit hard to swallow. It's not just about buying the tech but about everything that comes along with it; training, facilities, maintenance and updates. In some places, there's a struggle to get basic resources, let alone advanced AI tools.

Then there's the resistance to change. Some educators have been teaching a certain way for decades. Introducing them to AI is like handing them a recipe in a foreign language. They think, "Why fix what isn't broken?" Plus, there's a fear factor. What if AI replaces them? What if they can't keep up with the tech-savvy youngsters?

Of course, there are technology limitations. Not every place has high-speed internet or the latest gadgets. It's like wanting to cook a gourmet meal but only having a microwave. Some regions might struggle with power outages, outdated devices or a lack of tech support. The great thing is, change—though scary—can lead to growth!

Remember the kid and the broccoli? What if they end up loving it? What if it becomes their favorite veggie? Similarly, with the right approach, resources and mindset, the world of education can embrace AI one step at a time.

CASE STUDIES

I've always been a fan of stories. They say a lot more than plain facts.

1. The Rural School's Dilemma:

There's this school in a remote part of Kenya. They got their hands on an AI-based learning tool designed to personalize education for each student. Exciting, right? But they had limited internet connectivity. It's like having a Ferrari but no fuel. However, instead of giving up, they collaborated with the software providers to create an offline version. This resulted in students accessing lessons tailored just for them, even in the heart of the wilderness.

2. The Language Barrier:

In Thailand, a school wanted to use an AI tool developed in the US but there was a hiccup. The tool was in English and the students primarily spoke Thai. Instead of tossing the idea out the window, the school partnered with local tech enthusiasts. They adapted the tool, translating it and incorporating local cultural nuances. The outcome was an AI tool that felt homegrown, resonating with the students and making learning a breeze.

These stories remind me of something my grandma used to say, "Where there is a will, there is a way." These institutions faced challenges, yes, but with a dash of creativity and determination, they turned potential pitfalls into success stories that benefited many.

POTENTIAL PITFALLS:
THE DANGERS OF DEPENDING TOO HEAVILY ON TECHNOLOGY

It's fascinating how technology, especially AI, has woven its way into our lives, especially in education. But, like with all good things, there's a flip side. Picture a world where AI is the primary source of knowledge. Sounds convenient, right? But what happens when we lean too much on it?

I recently stumbled upon an article that highlighted this issue, focusing on the paradox of generative AI being both 'accessible' and 'restrictive'. OpenAI, for instance, launched tools like ChatGPT with the noble intention of democratizing knowledge.

This could be a game-changer for students facing language barriers, helping them overcome academic challenges but what happens when these tools suddenly become paid-only services after being freely available? This could inadvertently create a divide, favoring those who can afford them and leaving behind those who can't. It's like offering a taste of an elixir and then putting a price tag on it.

The question then becomes, should educational institutions bear the cost of these tools for their students? It's a tricky question. While Ivy League institutions might easily afford it, what about the smaller institutions in developing countries? The dream of a global educational equilibrium might remain a dream.

Some educational bodies quickly ban these tools, thinking it's the best way to control their use but banning them might make them more popular! It's like telling someone not to think of a pink elephant—now that's all they can think about. Instead of outright bans, maybe we should focus on integrating these tools into curriculums and promoting their ethical use. It's about guiding, not prohibiting.

LESSONS FROM FAILURES

You know how sometimes, despite your best efforts, your freshly baked cookies turn out a little....well, let's call them 'extra crispy'? Well, even with the best intentions, some AI initiatives in education didn't quite turn out as planned.

Imagine an institution decided to use an AI-powered tool to grade student essays. Sounds efficient, right? Not until the students realized the system could be tricked using fancy words, even if the content made zero sense. "The exuberant chicken danced flamboyantly under the cerulean sky." Grade A+! It's hilarious, sure, but it is not the outcome the institution hoped for.

Another institution believed they were making admissions fairer by using AI. Unfortunately, the system was trained on past admission data and therefore, perpetuated the same biases that existed before. Instead of leveling the playing field, it became a high-tech mirror of past prejudices.

To keep virtual classes engaging, a school adopted an AI tool that analyzed students' facial expressions to gauge their level of interest. Great in theory, but in practice? Not so much. Some students felt it was creepy, while others, like our poker-faced pals, were flagged as 'unengaged' even when they were paying full attention.

CONTINUOUS IMPROVEMENT

It's like kneading dough. It might be a bit sticky and not quite right the first time but it improves as you keep working on it and folding in feedback. Before you know it, you've got the perfect base for a delicious pizza (or, in our case, an AI tool that's on point).

Every tweak, every change, every "Aha!" moment is a step forward. It's about learning from experience, being receptive to feedback and being agile enough to pivot when needed.

Here's the best part—this cycle never really ends. There's always room to refine, enhance and elevate, because in the world of AI, just like in our bustling kitchen, standing still isn't an option.

KEY CHAPTER TAKEAWAYS

Resistance to Change: Humans are creatures of habit and introducing new technologies, especially something as transformative as AI, can be met with resistance. However, this resistance can be overcome with the right approach and mindset.

Potential Pitfalls: While AI offers numerous advantages, it's essential to be aware of potential pitfalls, such as biases in data or over-reliance on technology. It's crucial to approach AI critically, ensuring its integration is done ethically and responsibly.

The Power of Feedback: Continuous improvement is the key to successfully integrating AI into education. Feedback from all stakeholders, especially students and educators, is invaluable in refining and improving AI tools.

CHAPTER LEARNING ACTIVITIES

ACTIVITY 1: DEBATE SESSION

Organize a debate on the pitfalls and potentials of AI in Education. This will allow participants to delve deeper into the pros and cons, fostering a deeper understanding of the topic.

ACTIVITY 2: AI TOOL ANALYSIS

If you can access an AI educational tool or software, explore it. Note down its features, benefits and potential drawbacks. How would you improve it? This activity will give you a hands-on understanding of the current state of AI in education.

ACTIVITY 3: SCENARIO BUILDING

Imagine a classroom 20 years from now. How do you envision AI's role? Write a short essay or story describing a day in a student's life in this AI-integrated classroom. This will help you think creatively about the future possibilities of AI in education.

15

THE ROAD AHEAD: BEYOND AI

AI has already woven into education, revolutionizing how we teach, learn and interact but what if I told you that this is just the beginning? Beyond AI, a vast expanse of technological wonders is waiting to be explored. Bear with me; it's about to get exciting.

Imagine a world where the classroom has no walls, where students from Tokyo can dive deep into the Amazon rainforest, guided by a teacher in Toronto, all without leaving their homes. Virtual Reality (VR) and Augmented Reality (AR) make this possible. These aren't just "cool gadgets" anymore; they're shaping immersive learning experiences that transcend geographical boundaries.

Ever heard of Quantum Computing? I know it sounds like something straight out of a Star Trek episode. However, this technology has the potential to process complex problems at speeds previously deemed impossible. Imagine a world where educational research, currently taking years, is completed in mere seconds.

Now, let's talk about the brain. No, not in a philosophical, "What are we?" way. I'm talking about Brain-Computer Interfaces (BCIs). This cutting-edge tech can read our brainwaves, opening up a world of possibilities. For students with disabilities, BCIs could revolutionize accessibility, breaking down barriers and creating inclusive learning environments for all.

And let's not forget about nanotechnology. Imagine tiny robots, so small you can't see them, working away in the background, making our educational tools more efficient, durable and sustainable.

While these technologies might seem galaxies away, they're closer than you think. Pioneers in education are already experimenting, iterating and implementing these tools to build a brighter future for learners everywhere.

So, what's the takeaway? The road ahead is paved with possibilities. Beyond AI, a universe of technologies is waiting to be explored, each with the potential to redefine the very essence of education and you, my dear reader, are a part of this journey.

AR (Augmented Reality) is like that friend who brings an extra layer of pizzazz to your reality. Think Pokémon Go but educational. Imagine a biology class student pointing their tablets at a seemingly ordinary poster on the wall, only to see a 3D heart pulsating and coming to life right before their eyes. Their world remains unchanged, but AR adds digital elements, making learning more interactive and fun.

VR (Virtual Reality), on the other hand, is that magic portal that teleports you to a completely different world. Slap on those VR goggles and bam! You're standing amidst the ruins of Ancient Rome, feeling the cobblestone under your feet, hearing the distant chatter of the market, and the best part? No passport needed! It's immersive learning at its best, providing students with experiential knowledge that's hard to forget.

Now, I know what you're thinking. "Sounds fancy, but is it practical?" I hear you—integrating AR and VR into the classroom does require some investment, both in terms of money and training but the payoff is monumental. We're talking about increased engagement, improved retention and a sprinkle of that 'wow' factor that keeps students hooked.

Of course, there's a learning curve, as with all things tech. Teachers, bless their hearts, might feel a tad overwhelmed initially. Remember the first time you tried to set up your email? Yet, with a little patience and a pinch of training, you'll be able to navigate these tools like a pro, before you know it.

QUANTUM COMPUTING:
POTENTIAL IMPLICATIONS IN EDUCATION

You might have heard whispers of quantum computing being the 'next big thing', but let me tell you, it's more than just a fancy phrase. You might wonder, "What does this high-tech jargon have to do with education?"

Quantum computing isn't just a beefed-up version of our regular computers. Oh no, it's a whole different ball game. While traditional computers use bits (1s and 0s) to process data, quantum computers use qubits. These little wonders can be in a state of 1, 0, or both simultaneously (yeah, it's mind-boggling, I know). This means they can process information at lightning speed.

Now, let's hop over to education. Imagine the sheer volume of data schools and universities generate, from student performance metrics and online class interactions to behavioral patterns. Processing and analyzing this data mountain requires significant computational power and time. However, analyzing complex datasets could be as easy as pie with quantum computing and not just any pie, but the delicious, melt-in-your-mouth kind.

But hold on, it's not just about speed. Quantum computing could revolutionize personalized learning. Think about it: algorithms that adapt in real-time to a student's learning pace, predicting their needs and offering resources instantly. An ordinary computer might freeze for days with all those inputs, but not quantum! It's like a personal tutor that never needs a break—ever.

Okay, I'll admit, it's not all rainbows and sunshine. Integrating quantum computing into education will require massive investment, infrastructure overhaul and training. Teachers must embrace yet another technological shift and there will be the inevitable teething problems. Didn't we all grumble when smartphones first came out but now, can you imagine life without one?

THE HUMAN ELEMENT

While AI tools have the potential to democratize knowledge, however, these tools might not always be freely accessible to everyone, everywhere. That's where the human touch comes in.

Educators, with their innate ability to empathize and understand individual needs, play a pivotal role in bridging these gaps. They're not just knowledge dispensers but mentors, guides and sometimes even friends. Remember that teacher who stayed back after class to explain a concept you found tricky? Or the one who recognized your potential even when you didn't?

That's the irreplaceable human element. While AI can provide information, it can't provide the warmth, understanding and guidance a human educator can. It's like comparing a well-cooked homemade meal to a fast-food takeaway. Both might satiate your hunger, but only one nourishes your soul.

Generative AI tools, despite their potential, can be restrictive. They might be behind paywalls or have limited accessibility based on resources. And while they can provide vast amounts of information, they lack the human intuition and expertise educators bring.

What if you are learning to play the guitar—an AI can teach you chords, scales and even songs, but can it understand the subtle nuances of your playing style? Can it give you feedback on your strumming technique or the emotion you put into a song? Probably not, but a human teacher can. They can guide you, correct you and even inspire you.

EMOTIONAL INTELLIGENCE AND EMPATHY

No matter how advanced AI becomes, there are certain human qualities it just can't replicate. Emotional intelligence and empathy top that list.

Remember when you were in school and faced a challenging topic? It wasn't just the information from the books that helped but the understanding and patience of a teacher who sensed your struggle. That's emotional intelligence. It's the ability to perceive, understand and manage our own emotions while also being attuned to the emotions of others. Empathy is that warm feeling of someone genuinely understanding and sharing your feelings. It's the comforting pat on the back, the reassuring words and the shared joy or sorrow.

Now, don't get me wrong, AI is fantastic. It can process information quickly, offer solutions and even mimic human interactions but can it truly understand the subtle nuances of human emotions? Can it sense the hesitation in a student's voice, the unspoken fears or the hidden excitement? I doubt it. Because while AI operates on data and algorithms, humans operate on feelings, experiences and connections. That is something truly irreplaceable.

INTEGRATING AI WITHOUT LOSING THE HUMAN TOUCH

1. Complementary Roles: Think of AI as a tool, not a competitor. Just like how a calculator doesn't replace a mathematician but aids them, AI should be used to enhance human capabilities. It can handle data crunching while educators focus on the more nuanced aspects of teaching.

2. Open Discussions: Encourage open conversations about AI in classrooms. Let students, teachers and parents voice their opinions, share their concerns, and brainstorm ways to integrate AI beneficially.

3. Ethical Use: It's essential to set boundaries. While AI can assist in various tasks, there should be clear guidelines on what's acceptable. For instance, using AI to brainstorm ideas? Great! Using it to complete assignments verbatim? Not so much.

4. Continuous Learning: The world of AI is ever-evolving. Regular training sessions can help educators stay updated and utilize AI tools effectively.

5. Embrace the Human Element: Never forget the power of human interactions. While AI can provide information, educators provide context, mentorship and a safe environment for growth.

ENCOURAGING INNOVATION

Have you ever been in a room where everyone's so afraid to make a mistake you can practically feel the tension? Not fun, right? Now imagine a classroom, a school or an institution where mistakes aren't just accepted but celebrated! A place where 'failure' isn't a bad word but an opportunity to learn and grow. Sounds amazing!

To truly innovate, we need spaces where people can think outside the box and brainstorm wild ideas without fear of judgment. Remember when you had a wacky idea that turned out to be brilliant? That's the power of experimentation. Trust me, fostering such a culture isn't just about fancy gadgets and open-floor offices—it's about attitude. It's about saying, "Hey, it's okay to try and fail, because sometimes, when you try, you create miraculous things."

THE ROLE OF RESEARCH INSTITUTIONS

Aren't research institutions the unsung heroes of innovation? I mean, think about it. Behind every groundbreaking tech in education, a team of researchers probably spent sleepless nights working on it.

Research institutions are pivotal in pushing the boundaries of what's possible. They're like the backstage crew in a play, ensuring everything runs smoothly, though they might not always be in the spotlight, their work is essential! From devising novel teaching methodologies to developing state-of-the-art educational tools, these institutions are at the forefront of innovation but they can't do it alone. Collaboration is key. What if educators, students, parents and policymakers actively engaged with these institutions, sharing insights, challenges and needs. We'd have solutions tailored to real-world challenges.

TECH & EDUCATION: SECTOR COLLABORATION

Okay, so you know that age-old saying, "two heads are better than one"? Let's amplify that by a million when we talk about the collaboration between tech and education. On one side, we've got tech gurus dreaming up the latest gizmos and gadgets. On the flip side, we have passionate educators with the noble desire to disseminate knowledge to the masses. So, why not bring these two powerhouses together?

I once sat in on a meeting between tech developers and

teachers and the energy was electric! They brainstormed, debated and threw ideas around like confetti. The outcome? Solutions that neither could've imagined on their own. The real magic happens when these sectors genuinely understand and value each other's strengths.

KEY CHAPTER TAKEAWAYS

Beyond AI: The world of education is on the brink of even more technological advancements, from Virtual Reality (VR) and Augmented Reality (AR) to Quantum Computing and Brain-Computer Interfaces (BCIs).

The Human Touch: No matter how advanced technology becomes, the irreplaceable value of human interaction, empathy and emotional intelligence remains paramount in education.

Collaboration is Key: The future of education lies in the collaboration between tech developers, educators, research institutions and other stakeholders.

CHAPTER LEARNING ACTIVITIES

ACTIVITY 1: VIRTUAL EXPLORATION

If you can access a VR headset or AR-enabled device, explore an educational application. Reflect on your experience. How did it differ from traditional learning methods? What are the potential benefits and drawbacks?

ACTIVITY 2: BCI EXPLORATION

Research a real-world Brain-Computer Interface (BCI) application in education. How do you see this technology evolving in the next decade? How might it revolutionize classrooms, especially for students with disabilities?

ACTIVITY 3: DISCUSSION FORUM

Engage in a discussion with peers or colleagues about the ethical implications of integrating advanced technologies like Quantum Computing in education. Consider questions like: Who should have access? How can we ensure equity? What potential biases should we be aware of?

14

AUGMENTED REALITY (AR) AND VIRTUAL REALITY (VR) IN EDUCATION

THE IMMERSIVE LEARNING EXPERIENCE

I've always been a believer in the magic of stories. The kind that sweeps you off your feet, makes you forget the real world and transports you to limitless imagination. So, what if I told you we can harness this enchantment for education? Yep, that's where AR and VR come in.

Imagine studying marine life by diving deep into the ocean without getting wet or testing out new advanced surgery techniques without needing a living, breathing person as your guinea pig. Gone are the days when you need to sit in a traditional classroom, squinting at a blurry map of Ancient Rome, trying to image those magnificent gladiatorial games, the roar of the crowd or the adrenaline. Now, you can strap on a VR headset and suddenly, you're there. You can hear the crowd, see the vibrant city and feel the pulsating energy. Which experience do you think you'd remember more vividly?

Immersive learning isn't just about fancy graphics or making lessons 'fun'. At its core, it taps into our brain's innate ability to remember experiences better than facts. By living an experience, we're not just passive recipients of information; we become part of the narrative.

From a psychological perspective, immersive environments engage our senses in a manner traditional methods can't. When multiple senses are engaged, our brains form stronger memories. It's the difference between reading about a concept and living it.

Moreover, for our visual learners out there, AR and VR are a godsend. They transform abstract concepts into tangible experiences. Struggling with geometry? Why not walk inside a 3D shape? Finding history boring? Time-travel and attend a lecture by Abraham Lincoln!

AR VS. VR

Think of AR as the world you know but sprinkled with a bit of digital magic. Have you ever used a filter on social media that gives you dog ears or dancing unicorns around your head? That's AR in its most playful form. It overlays digital elements in the real world, making reality more "augmented." Imagine teaching geography with 3D landmarks popping up from textbooks for our educators. Engaging, right?

VR takes things deeper however. Instead of merely augmenting our world, VR says, "Hold my headset!" and plunges you into entirely new worlds. It's immersive to the point where you can find yourself in the vastness of space or the ocean's depths with a headset. A teacher could transport their students to the

Renaissance era, making history lessons feel like time travel adventures!

While both AR and VR are champions in their own right, their applications differ. AR enhances our world, making it an incredible tool for interactive learning, hands-on training and enhanced user experiences. Conversely, VR offers full-blown simulations, perfect for immersive educational experiences, advanced training modules or therapeutic interventions.

I remember the first time I slipped on a VR headset—I was half-expecting to be transported into the Matrix. Instead, I found myself swimming with dolphins. This was made possible with a mix of hardware and software applications.

Now when I say "hardware", think of it as the physical instruments you'd touch or wear. With VR, it's those chunky headsets like the Oculus Rift, Vision Pro or the HTC Vive. This hardware becomes your gateway into another world. For AR, you have things like Microsoft's HoloLens, which overlays holograms onto your environment but what's a great piece of hardware without its trusty software sidekick?

Enter the software platforms. Platforms like Unity or Unreal Engine make those fantastical worlds alive for VR. For AR, Apple's ARKit and Google's ARCore are leading the parade, allowing developers to anchor digital objects in the real world. It's like giving a painter a canvas and a three-dimensional world to play with. With these tools, High School Biology isn't just diagrams anymore, it's "walking" inside a human cell and it's all thanks to these incredible pieces of hardware and software.

VIRTUAL LABS AND FIELD TRIPS

I remember "field trips" meant packing a lunch, hopping on a bus and touring a local museum or factory. Don't get me wrong; I loved those. The smell of the museum, the buzz of my classmates and the thrill of the unexpected. Times have changed my friends and technology opens doors to places we never thought possible.

With VR and AR, we're no longer limited to our textbooks or the local science center. We can now explore in a world where boundaries don't exist. Students in New York can explore the jungles of the Amazon, while those in Tokyo can explore the cobbled streets of Rome. All of this without the jet lag or the expenses! Virtual experiences are redefining how we perceive education. They're adding layers, textures, and, most importantly, experiences that textbooks can't replicate.

Now, let's talk about the labs. Remember those Bunsen burners and the occasional experiment gone wrong? Well, with virtual labs, things have taken a turn for the better. No longer confined by physical limitations or safety concerns, students can now conduct experiments that were once deemed too complex or hazardous. The beauty of this? Mistakes are okay; they're encouraged even! After all, in a virtual world, there's always a reset button.

This is not just about the wow factor. It's about access and equity. Not every school can afford state-of-the-art labs or is located near world-class museums. Virtual experiences level the playing field. Regardless of residence, every student gets a front-row seat to the best educational experiences.

HOW VIRTUAL LABS CAN OFFER HANDS-ON EXPERIENCES

We live in an era where 'virtual' is the new 'real'. This shift is nothing short of magical in science subjects like chemistry, physics and biology.

Virtual labs are a game changer. They allow us to dissect a frog without the... umm... squishy bits or play with chemical compounds, minus the fear of an unexpected explosion (unless that's the lesson for the day, of course!). These nifty tools simulate environments that are too complex, dangerous or impossible in the real world.

For instance, budding biologists can dive into the human body, from the adrenaline-pumping heart chambers to the intricate neural networks. Physics enthusiasts? They can launch rockets, mess around with quantum mechanics and even toy with relativity. All from the safety and comfort of their classroom or homes.

Amazing, right? It's not just about the cool factor, however, it's about diving deep, getting a feel of the subject and fostering a genuine love for learning. I wish I had this when trying to decipher the mysteries of DNA or the wonders of electromagnetic waves!

THE BENEFITS OF VIRTUAL FIELD TRIPS

I've always been fascinated by the Pyramids of Egypt but the closest I ever got was a dog-eared textbook and a half-remembered documentary. Enter virtual field trips. Now, students can stand at the base of these ancient wonders, look up and almost feel the desert wind on their faces.

Subjects like history, geography and cultural studies burst into life when experienced up close. The Great Wall of China isn't just a line on a map; it's a sprawling structure with tales etched into every brick; and cultural festivals? Watching them virtually means not just seeing but experiencing the rhythm, colors and life of faraway lands.

By virtually wandering the globe, students foster a sense of global citizenship. They connect the dots, empathize with diverse communities and grasp the richness of our shared human heritage. The best part? No permission slips or packed lunches are needed!

CHALLENGES AND SOLUTIONS IN VIRTUAL EXPLORATION

1. The Cost Factor: Not every school has a budget that stretches like a rubber band. VR gear and software is cheaper than taking a group of students on a exploration field trip around the world, but let's face it, it can still be pricey.

- **Grants and Funding:** Before you say, "Not another grant application," hear me out. Tech companies, understanding the educational value of AR and VR, often have grant programs. So, a quick search and a well-crafted proposal might have those dollars rolling in and don't even get me started on the countless educational non-profits eager to support such initiatives.

- **Shared Resources:** Have you ever been to a potluck dinner? Everyone brings a dish, and suddenly you have a feast! Similarly, why not have a tech potluck? Schools in a district can pool resources, buy equipment and then rotate its use. Think of it as a merry-go-round of tech!

- **DIY Solutions:** Now, this might sound a tad out there, but with a little creativity, you'd be surprised what you can achieve. Cardboard VR viewers, open-source AR apps and even student-led projects to develop AR/VR content can significantly cut costs.
- **Rent Instead of Buy:** Remember when you borrowed that fancy suit for your cousin's wedding? (Yeah, you still need to return that). The same principle applies here. Some companies offer rentable AR and VR kits tailored for educational purposes.

2. Accessibility: It's one thing to have a single VR device and entirely another to ensure every eager young learner gets a turn. However, with cloud-based VR platforms and collaborative tools, multiple students can explore virtual journeys simultaneously. Plus, virtual platforms are increasingly focusing on being user-friendly for differently-abled students.

3. Reality gap: While VR can simulate the Roman Colosseum's grandeur, can it capture the actual atmosphere, the chirping of the birds and the distant chatter? Maybe not entirely. With enhancements in haptic feedback and 3D audio, the virtual world feels more tangible than ever.

4. Hardware: Have you ever tried watching a 4K video on an old computer? Yeah, that's a recipe for frustration. The world of AR and VR is no different. Using outdated devices can be like trying to race in an old, beat-up car. The solution? Always ensure your devices meet the minimum requirements, and if they don't, maybe it's time to organize a little fundraiser.

5. Software: You've got the gear, but what about the software? Compatibility issues can throw a wrench in the works. Regularly update software and check if the AR/VR program is compatible with your device, otherwise you might pull out some hair in the process.

6. Connectivity: A slow or unreliable connection can damage your virtual escapades. I remember trying to stream my favorite movie and the buffering circle became the main character! Invest in a stable internet connection, and consider wired options for crucial sessions.

PEDAGOGICAL ADJUSTMENTS

Remember the first time you tried sushi? Maybe it felt odd, but now you're all about that wasabi life. Similarly, stepping into AR and VR might seem alien, but it becomes second nature with time.

Training Programs: It's not just about having the tools but knowing how to use them. Investing in comprehensive training sessions can be a game-changer. Attend workshops, online courses, and even some fun AR/VR game nights to get the hang of it.

Mentorship and Collaboration: Buddy up with tech-savvy colleagues or even students. They can offer fresh perspectives, tricks and solutions. It's like teaming up in a cooking challenge—one chops the veggies, the other stirs the soup. Together, you create a mouthwatering dish; or, in our case, an unforgettable learning experience.

Feedback Loops: Once you start integrating AR and VR, gather feedback. What worked? What didn't? Was it like watching a 3D movie without the glasses (blurry and confusing) or more Avatar-esque (immersive and magical)? Adjust, adapt, and try again.

KEY CHAPTER TAKEAWAYS

Immersive Learning: AR and VR offer unparalleled immersive learning experiences, making abstract concepts tangible and memorable.

Benefits Beyond the 'Wow' Factor: These technologies aren't just about flashy graphics; they tap into our brain's ability to remember experiences better than facts.

Challenges and Solutions: While AR and VR are transformative, they come with challenges like cost, accessibility and the need for updated hardware and software. However, many of these challenges can be addressed with creativity and collaboration.

CHAPTER LEARNING ACTIVITIES

ACTIVITY 1: VIRTUAL SCAVENGER HUNT

Using a VR platform of your choice and embark on a virtual field trip. Create a list of items or landmarks to find during your journey. Share your findings with peers or in a classroom setting.

ACTIVITY 2: AR CLASSROOM ENHANCEMENT

If you're an educator, pick a topic from your curriculum and brainstorm how AR could enhance that lesson. For instance, if you're teaching about the solar system, could students use AR to "see" the planets in 3D around the classroom? Design a brief lesson plan around this idea.

ACTIVITY 3: TECH DEBATE

Organize a discussion or debate on the ethical implications of AR and VR in education. Consider questions like: How do we ensure equal access to these technologies? What are the potential drawbacks of relying too heavily on virtual experiences?

15
THE METAVERSE AND EDUCATION

Picture the 80s—neon lights, groovy music, and...the birth of the Metaverse idea! It was a time when most of us were jamming to Madonna and trying to solve that confounded Rubik's Cube, yet, in the shadows of popular culture, the concept of the Metaverse was silently brewing.

When I say "Metaverse," you might think of recent tech innovations but would you believe me if I said it all started with fiction? Our beloved authors dreamt up vast, interconnected virtual universes before we even had the internet in most households. Remember the novel "Snow Crash" by Neal Stephenson? The protagonist, Hiro, dabbled in a virtual world, an early representation of our current Metaverse ambitions.

Fast forward to the late 90s and early 2000s. We witnessed the rise of online games and virtual worlds. Ever heard of 'Second Life'? It wasn't just a game but a glimpse into a parallel universe where you could be whoever and do whatever. A prototype Metaverse, if you will.

Now, here's where things get extra spicy. With AR, VR and AI advancements, the lines blur between fiction and reality. Giant tech companies, indie developers, and even Aunt Gertrude's tech-savvy grandson are all jumping on this train, contributing their visions and innovations.

What started as a spark in an author's mind has become a blazing trail, leading us into an era where education, social interactions, work and play could occur in expansive virtual worlds.

COMPONENTS OF THE METAVERSE

1. Virtual Worlds: Imagine a place where you can sip a mocha latte on Mars or groove to alien beats in a nightclub on Pluto. Virtual worlds are these vast, immersive worlds where your imagination is the only limit. Think 'Minecraft' or 'Fortnite,' but dialed up to 100! Between you and me, I tried building a replica of the Eiffel Tower once—let's say it didn't end well for my virtual neighbors.

2. Digital Avatars: Imagine you're in one of these virtual worlds, and instead of being...well, you, you're a fire-breathing dragon, a rockstar or even a slice of animated pizza. These are digital avatars! They're our digital twins, our very essence in pixel form. Fun fact: My avatar has a pet llama named Larry. Don't ask why; it's a long story.

3. Interconnected Digital Spaces: Ever wished you could jump from a virtual concert in one game to a space battle in another, all in a single leap? In the Metaverse, you totally can! These spaces are all linked, creating an expansive, endless universe of experiences.

NAVIGATING THE METAVERSE ECOSYSTEM

Have you ever strolled through an app store and seen a bunch of VR or AR apps? That's them! These platforms host the virtual worlds we just chatted about. I remember when I almost flooded a virtual city trying to build a swimming pool? Yep, did that on one of these platforms. (Note to self: always hire a virtual architect).

Think of them as the stagehands in a grand theatre play. These technologies, from VR headsets to haptic suits, make our Metaverse experiences feel oh-so-real. By the way, have you ever tried dancing with a VR headset on? Word of advice: make sure you're alone. Trust me.

The Movers and Shakers: Like any city, the Metaverse has bigwigs. They're the creators, investors and innovators pushing boundaries every day. And honestly, some of their ideas? Out of this world! Literally.

Have you ever imagined a virtual school in the Metaverse? Picture having Einstein as your physics teacher or Shakespeare for drama class. How cool would that be?

Remember those coins you collect in video games? Now, they've grown up and gotten a degree in economics. Virtual currencies like Bitcoin and Ether are no longer just fun tokens; they drive real trade in the Metaverse. The way we socialize? It's going through a cosmic shift! Imagine going to a concert, hanging out with pals and even attending a poetry reading without leaving your room.

LEARNING IN A METAVERSE

When I was a kid, classrooms meant four walls, a chalkboard and wooden desks. Fast forward to today and the digital transformation has us envisioning education on an entirely different plane—from virtual labs to libraries without limits, the Metaverse offers tools and environments that we could only dream of. Last week, I wandered through a digital art gallery where each painting was a history lesson. Van Gogh's "Starry Night"? A gateway to astronomy!

Interacting with fellow learners and teachers in real-time, in a setting of your choice (hello, underwater biology class), enhances engagement. The Metaverse makes passive learning a thing of the past. Trust me, when you're debating Shakespeare with an AI or discussing math atop a virtual mountain, it's an experience, not just education.

EDUCATION IN A FULLY IMMERSIVE DIGITAL ENVIRONMENT

Think of it as Hogwarts, but designed by you. It's a space where the magic of learning is only limited by imagination. Why listen to a lecture on volcanoes when you can stand next to one (safely!) and watch it erupt? The Metaverse is like this colossal playground where lessons come to life.

From exploring the ocean's depths to wandering the Saharan deserts, geography got much more hands-on and for my math lovers, how about navigating through complex 3D geometric shapes? Suddenly, calculus feels like an adventure.

For teachers, envision crafting lessons where students aren't just passive listeners but active participants. The role of educators evolved from mere information dispensers to experienced curators. By the way, I once attended a virtual literature class inside a dragon's lair. Talk about setting the mood for a Beowulf reading!

Lastly, remember that awkwardness of the first day at school, finding your place, figuring out where to sit? In the virtual classroom, each student gets a front-row seat. Accessibility, inclusivity and personalization? Check, check and check.

COLLABORATIVE LEARNING IN THE METAVERSE

Remember when group projects meant huddling in someone's living room, frantically trying to assemble a presentation or a model? Well, times have changed! The Metaverse is here, giving "teamwork" a whole new dimension (literally!). Picture brainstorming with classmates on a floating platform above the clouds or collaborating on a physics experiment in zero gravity. Sounds like a sci-fi movie?

But it's not all fun and games (although there's plenty of that!). The Metaverse breaks down geographical restrictions, meaning Juan from Mexico can seamlessly work on a history project with Aisha from Nigeria and Chen from China. We're talking real-time, folks! No more time zone hassles or connectivity issues. The world just became your classroom and every student is your potential group mate.

What's more, the depth of interaction is unparalleled. Remember those heated classroom debates? They've been amplified! Now, you're not just discussing the Amazon rainforest; you're IN the Amazon, feeling the humidity, hearing the sounds, all while passionately arguing about its significance.

CHALLENGES AND OPPORTUNITIES

"It's all fun and games until you realize not every kid has access to the same toys." Isn't that a thought? While the Metaverse promises wonders, there's this looming cloud called the 'digital divide'. Not everyone has high-speed internet or the latest AR goggles. It's not just about gadgets. It's about opportunities, about potential and about dreams.

With challenges come chances to innovate, right? Education institutions, tech gurus, and policymakers—I'm looking at you. Can we find ways to bridge this divide? Scholarships, community tech hubs, maybe even some out-of-the-box solutions?

And while we're on this roller coaster of thoughts, let's not forget the goldmine of enhanced learning outcomes. When done right, the Metaverse can be the key to unlocking untapped potential in students. A space where visual learners, auditory learners and kinesthetic learners all find their groove.

METAVERSE ETHICS AND PRIVACY

DIGITAL IDENTITY AND PRIVACY

Do you remember the thrill of getting your first library card or maybe your driving license? The notion of identity has always been pivotal, but this takes on a new dimension in the Metaverse. Our students, from tiny tots to almost adults, are navigating this colossal digital expanse and crafting their digital identities. Sounds fun, right? Until you realize that every click and interaction is recorded.

I overheard a teenager say, "It's just online; it's not real." That hit me. The line between the virtual and the real is blurring, but the implications of our online actions are as real as they come. As such, students' privacy in such spaces is paramount. Imagine creating a virtual 'privacy shield' or running digital literacy programs focusing on the importance of online discretion. This are things we must not just talk about but implement.

ISSUES OF MISINFORMATION, DIGITAL ADDICTION AND ONLINE BEHAVIOR

The Metaverse is not just unicorns and rainbows. It's a melting pot of information; not all golden. It's like that sneaky chocolate wrapper you accidentally step on in a silent library—unexpected, disruptive and oh-so-noticeable.

Remember those late-night binge-watching sessions? The "just one more episode" promise we all break? Digital addiction in the Metaverse is that but on steroids. Students can get so enamored by this digital wonderland that they forget there's a world outside. We must strike a balance with the right guidance and perhaps some tech breaks.

Then comes online behavior. Have you ever come across those keyboard warriors? The ones hiding behind screens, throwing around words without realizing their weight? That's an aspect of the Metaverse we need to address. It's about creating a culture of empathy, understanding and kindness. That change starts with us.

SAFEGUARDING DIGITAL INTERACTIONS

Imagine you're exploring this vast digital universe, the Metaverse, and then —BAM!—You bump into a troll. Not the cute, glittery kind. The nasty, spoils-your-day kind. The Metaverse can be a double-edged sword, especially for our young learners. On one hand, there's endless potential for creativity, learning and fun. On the flip side, running into less-than-pleasant entities is possible.

Here's where I had an "aha!" moment: why not design 'safety zones'? Think of them like those safe play areas for toddlers, but digital. A space where students can interact, learn and collaborate without fearing unwanted disturbances. And while we're dreaming big, how about some robust AI moderators to keep things in check? But remember, it's not just about building barriers; it's about fostering respect, empathy and digital etiquette. It's like teaching kids to look both ways before crossing the street, but for the Metaverse.

FUTURE PROSPECTS

Where's the Metaverse headed and what does it mean for education? What if schools as we know them become obsolete? Imagine an expansive digital university where everyone, irrespective of geography, has access. The democratization of education! Though I'd miss the old-school chalkboards.

There are challenges, of course. Ensuring equitable access, battling the digital divide, and maintaining the human touch in teaching but the possibilities are there and the potential is exhilarating.

KEY CHAPTER TAKEAWAYS

Origins and Evolution: The Metaverse, while a buzzword today, has its roots in fiction from the 80s, evolving through online games and virtual worlds of the 90s and 2000s.

Components and Ecosystem: The Metaverse is a blend of virtual worlds, digital avatars and interconnected digital spaces powered by platforms, technology and stakeholders.

Learning in the Metaverse: Beyond traditional classrooms, the Metaverse offers immersive, experiential learning opportunities, breaking geographical boundaries and redefining collaboration.

Challenges and Opportunities: While the Metaverse promises revolutionary educational experiences, it also brings forth challenges like the digital divide, privacy concerns, and the need for digital etiquette.

CHAPTER LEARNING ACTIVITIES

ACTIVITY 1: CRAFT YOUR DIGITAL IDENTITY

Dive into a Metaverse platform of your choice. Create a digital avatar that represents you. Reflect on the choices you make. What do they say about your identity, aspirations or personality?

ACTIVITY 2: VIRTUAL FIELD TRIP

Join a virtual world or an educational platform within the Metaverse. It could be a historical site, a scientific phenomenon or an art gallery. Document your experiences, insights and learnings. How was this different from reading about it in a textbook?

ACTIVITY 3: DEBATE AND DISCUSS

Organize a virtual discussion or debate within a Metaverse platform on the topic: "The Metaverse in Education: Possibilities and Pitfalls?" Encourage participants to discuss the potential, challenges and ethical considerations.

16
BLOCKCHAIN TECHNOLOGY IN EDUCATION

Let's say you are the boss and you're hiring for a new role, two candidates walk in, claiming to have graduated from Harvard. But wait a minute! There's no way to verify this easily. Enter Blockchain, the hero of transparency and credential verification in education.

When I first heard about Blockchain, my brain did a little "what...?" But trust me, it's a simple and elegant concept once you get past the jargon.

Blockchain is the equivalent of a digital ledger or endless digital book. Every time something noteworthy happens, like you earning a diploma or achieving a certification, a new page gets added to this book. Now, this isn't just any ordinary book. Each page, once written, is nearly impossible to change. Copies of this book are spread out everywhere, so tampering with one would mean tampering with all.

In simpler terms, Blockchain is a chain of blocks containing information, like who owns what or who did what. A new block is formed every time an older block is filled up with info and each block contains a tiny piece of the previous block. It's like a digital version of those Russian nesting dolls.

So, how does this relate to education? Imagine if every time a student earned a certificate or completed a course, it was recorded on this Blockchain. A digital badge, if you will. This would make the credentials tamper-proof and easily verifiable. No more fake degrees, folks! It's like a breath of fresh air for educators and institutions that have been grappling with the issue of credential fraud for ages.

DATA INTEGRITY AND SECURITY WITH BLOCKCHAIN

Have you ever kept a diary? You know, one of those where you jot down your deepest secrets and desires. Imagine if that diary was out there for everyone to see, but with a magical lock and key ensuring no one could alter your words. Fascinating, right? This is where the beauty of the Blockchain's 'immutable ledger' comes into play.

The term 'immutable ledger' may sound like something out of a sci-fi novel, but it's pretty straightforward. 'Immutable' means unchanging, and 'ledger' is a fancy word for a record or a list. Put them together, and you've got an unchangeable record.

Imagine you're playing a game of 'pass the parcel'. But in this version, every time the parcel reaches a player, they put a tiny note inside without disturbing the previous notes. Blockchain is like that game. Each time data gets added (think of it as the new note), it's linked to the previous one. This chain ensures that it stays safe and secure once the data enters, just like the notes in our parcel.

You maybe asking why can't someone...you know...change the stuff inside the parcel? Here's the cool part: each note (or data block) has a unique digital fingerprint code. This code is influenced by the note before it. So, any attempt to mess with a note would mean changing every note after it.

This structure not only keeps our data safe but also transparent. Everyone can see the notes, but they can't alter them. It's a game-changer, especially in sectors like education where every certificate, every grade and every achievement is recorded. The possibilities are endless and exciting!

I remember back in school, I'd sometimes scribble in the margins of my books. These little notes were my insights, ideas or just random doodles. The Blockchain is like the main text, while the little scribbles are the multiple applications it offers, from cryptocurrencies like Bitcoin to credential verification in education.

DECENTRALIZED LEARNING RECORDS

You know those special moments in movies where the protagonist discovers they have the power to control their destiny? That spine-tingling realization that they, and they alone, can shape their future? What if I told you that we are on the cusp of a similar revolution in education? A shift that's about to put students at the heart of their academic journey, giving them unparalleled control. Sounds like a blockbuster plot.

Enter the world of decentralized learning records. A mouthful, I know, but stick with me, and I promise you'll be as excited as I am by the end.

Let's start with a fun little exercise. Think of your academic records as a treasured family heirloom passed down through generations. It's precious and personal and carries the legacy of your hard work. Would you hand it over to a distant cousin thrice removed to safeguard it? I didn't think so! In a sense, that's what we've been doing with our educational data. Schools, colleges and even online platforms hold onto our records. We trust them, often blindly, but what if they lose it or, even worse? Scary thought, right?

The age-old system where institutions are the custodians of our records is getting a modern makeover and it's about time! With the rise of Blockchain and decentralized technologies, the power is shifting back to the students.

Picture a world where every assignment you've submitted and every test you've aced (or maybe not-so-aced, but hey, we've all been there) is stored safely in a digital vault, and you hold the key! Only you can decide who gets a peek inside. Want to show a potential employer your stellar grades? Go right ahead. Thinking of sharing your progress with a mentor? Your call. The point is, you're in the driver's seat.

You might wonder, "Why is this such a big deal?" Well, think about it. When you have control over your data, you're not just a passive bystander but an active participant in your educational journey. You decide your path, and you shape your narrative.

There's also this added layer of security and authenticity. No more worries about records getting misplaced or tampered with. We've all heard those horror stories of credentials getting fudged. With this new system, that's all set to change.

UNIVERSAL EDUCATION WITH THE IMMUTABLE LEDGER

Imagine you're on a worldwide culinary tour. Today, you're relishing sushi in Tokyo; tomorrow, you're sipping on espresso in a quaint Roman café. Every time you hop from one place to another, you need a new currency. It'd be much easier if there were a universal currency, right? Let's swap out these cuisines and currencies for academic records and institutions. That's where Blockchain and interoperability come into play.

Ever faced the hassle of transferring academic records from one institution to another? It's like trying to fit a round peg into a square hole. Every institution has its system, its format and its little quirks. Enter Blockchain; it's like the superhero we didn't know we needed.

With Blockchain, records are stored in a universally accepted format. Think of it as the educational world's Esperanto—a language everyone understands. So, when Jane wants to transfer her credits from a university in Canada to one in Australia, it's no longer a paperwork nightmare. The records are effortlessly recognized and accepted.

This isn't just about saving time or cutting red tape. It's a game-changer for ensuring that education is accessible and continuous. Students can now weave together diverse learning experiences from across the globe. One month, they study art in Paris; the next, they dive into marine biology in the Maldives.

LIFELONG LEARNING PORTFOLIOS

Image your favorite bookshelf, the one that's witnessed your life story. Instead of books, imagine each spine is a different experience or skill you've gained over the years, from that summer pottery class to your master's degree, each one holds a special memory. This is what we envision when we talk about Lifelong Learning Portfolios.

It's more intricate than that artsy bullet journal you started last year, but bear with me. These portfolios aren't just a list of certificates or degrees. Oh no, they're a vibrant mosaic of formal education and the random skills you've picked up on lazy Sunday afternoons. Remember that YouTube rabbit hole you fell into learning about the art of Bonsai?

Think of it as your own personal IMDb page. Instead of movies and TV shows, it's filled with workshops, online courses, internships and even those dance lessons you took when you were five. Every experience, no matter how small, has shaped you in some way, and it deserves its rightful place on that shelf.

I can't help but get giddy thinking about how transformative this would be. Imagine applying for a job or joining a new course and instead of the usual CV, you hand over this colorful tapestry of your life's learnings. It tells your story, showcases your versatility and highlights the unconventional, making you stand out.

CHALLENGES AND SOLUTIONS

Standardization: Imagine trying to make a puzzle, but each piece is from a different set. That's what we're grappling with here. Every institution has its way of doing things. Some prefer quills and ink, while others are all about that futuristic holographic tech (well, maybe not, but you get the point). So, we're looking at creating a one-size-fits-all standard which is easier said than done—but not impossible. A universally accepted protocol might be the answer. It's like deciding on a universal handshake, perhaps the first bump? Kidding! But you catch my drift.

Adoption: Remember when we shifted from chalkboards to whiteboards, and Mr. Thompson couldn't find the eraser? (Sorry, Mr. Thompson) Change can be tricky. However, with proper training and a tad bit of patience, we can smooth out these bumps. Trust me, the result will be worth it.

Privacy: We live in an age where data breaches are more common than my aunt's 'famous' casserole at family gatherings. Protecting sensitive data is crucial. Using encrypted platforms and rigorous security measures can be our shield in this data-driven gladiator arena.

SMART CONTRACTS

Have you ever wished for a contract that doesn't need lawyers or notaries but...works? Let me introduce you to the world of smart contracts. Imagine you and your friend, let's call him Bob, betting on whether it'll rain tomorrow. You both put $10 into a box, and there's a little elf inside (stay with me here!) who checks the weather. If it rains, the elf hands Bob the $20. If not, you get it. Simple, right?

Well, smart contracts are just like that, minus the elf. Instead of an elf, we've got lines of code. When certain conditions are met (like rain or no rain), the contract self-executes and appropriate action is taken. It's automatic, trustworthy and devoid of any human intervention.

MICRO-CREDENTIALING

Have you ever been in a situation where you've acquired a specific skill—say, baking the perfect brownie—but had no way to show it off officially? Enter: micro-credentialing.

In the educational world, not all learning fits neatly into a three-year degree or a diploma. Sometimes, those small, specific skills or bite-sized chunks of knowledge truly matter. Micro-credentialing is like giving a high-five or a gold star sticker for these mini achievements. Think of them as badges of honor for the modern learner.

This is about recognizing that sometimes, it's not the massive degrees but the tiny, specific skill sets that can make all the difference. Maybe it's proficiency in a software tool, mastering a craft technique or even understanding the nuances of a niche subject. Whatever it is, micro-credentialing says, "Hey, we see you and that awesome thing you learned. Here's your badge of honor!"

THE FUTURE OF CERTIFICATION

Imagine you've just finished a riveting course on, let's say, the art of making the world's best sandwich (because who doesn't love sandwiches?). Instead of a piece of paper that could get lost or damaged, you receive a digital certificate stored on a Blockchain. A digital proof that you are a sandwich-making maestro. This is the future of certification.

Blockchain is like that super-organized friend who keeps a meticulous diary. Every entry, once made, can't be tampered with. Similarly, once your learning is recorded on the Blockchain, it's there for good—safe, secure and verifiable by anyone. No more lost certificates or doubts about authenticity. The future seems promising. Sandwich-making aside, this could revolutionize how we recognize and validate all forms of learning.

ETHICAL AND PRACTICAL CONSIDERATIONS

Just because we can do something, should we always jump right in? With great power (or technology) comes great responsibility. The use of Blockchain in education sounds glitzy and glamorous, but we must tread with caution. We must ensure it's accessible to all, not just the tech-savvy. Fairness is paramount. Can you imagine being left out because you're not 'in the know'?

Moreover, we should use blockchain responsibly, ensuring data privacy and understanding the environmental implications it poses. Just like traditional industries leave behind waste and harmful byproducts of their use, blockchain has similar issues. If you thought light bulbs were energy-intensive, consider the implications of a global processing unit. With new technology come new concerns, such as electronic waste, carbon footprints and much more.

KEY CHAPTER TAKEAWAYS

Transparency and Credential Verification: Blockchain offers a tamper-proof way to verify educational credentials, ensuring authenticity and trustworthiness.

Decentralized Learning Records: Students are at the center of their academic journey, holding the key to their educational data.

Smart Contracts and Micro-Credentials: Recognizing specific skills and knowledge chunks and automating processes in education.

Ethical and Practical Considerations: With the potential of Blockchain comes the responsibility of ensuring fairness, accessibility and sustainability.

CHAPTER LEARNING ACTIVITIES

ACTIVITY 1: DIGITAL BADGE DESIGN

Imagine you've just completed a course on Blockchain basics. Design a digital badge that you would want to receive. What elements would it include? How would it represent the course's content?

ACTIVITY 2: DEBATE ON DECENTRALIZATION

Organize a debate or discussion group focusing on the pros and cons of decentralized learning records. How might they change the landscape of education? What challenges might arise?

ACTIVITY 3: ETHICAL DILEMMA

Consider the environmental implications of Blockchain, given its energy-intensive nature. Write a short essay or reflection on how educational institutions can balance the benefits of Blockchain with environmental responsibility.

17

INTEGRATING AR, VR, BLOCKCHAIN, AND AI IN EDUCATION

COMPLEMENTING TECH IN EDUCATIONAL SETTINGS

So far we have been talking about AR, VR, Blockchain and AI as isolated entities within education but have you ever thought what might happen if we mix them all together? Take that history lesson on ancient Rome. With AR (Augmented Reality), students can overlay 3D images of historical artifacts onto their real-world environment. So, instead of just reading about a gladiator's helmet, they can virtually "hold" and inspect it in their hands. Once done, a student can use VR (Virtual Reality) to walk the actual gladiator's arena and even watch the battle but what if the student missed something?

While AR and VR offer immersive experiences, they still rely on vast data and that's where AI (Artificial Intelligence) comes into play. Picture these immersive experiences specially customized for each student, adapting in real-time to their needs. For instance, if a student struggles with a concept, the AI-driven system provides additional resources or changes the virtual scenario to aid understanding.

You might wonder, "This all sounds awesome, but how do we ensure that the data and certifications from these experiences are secure and verifiable?" Enter Blockchain. Remember that nifty digital certificate we talked about earlier? Blockchain ensures that any achievements or credentials obtained in these virtual spaces remains tamper-proof and verifiable. Now students can proudly showcase their accomplishments without any worry.

The real beauty lies in how all these technologies amplify each other's strengths. AR and VR offer immersive, hands-on experiences that make learning memorable. AI tailors the educational journey to the individual, ensuring no one's left behind. And Blockchain? Well, it's the trusty guardian keeping everything in check.

INTEGRATED LEARNING ENVIRONMENTS

Imagine little Timmy entering the room. He puts on a sleek pair of AR glasses, and suddenly, those pesky math formulas become floating holograms. He can grab, rotate and even interact with them. Who said math wasn't fun? Remember those heavy backpacks that felt like you were carrying a small elephant? Not here! Everything's digitized and personalized. Textbooks, assignments, notes...all accessible with a simple gesture or voice command. Believe me, your back will thank us later.

While Timmy's navigating the wonders of quadratic equations in 3D, Maria is on a VR field trip. Where? The Swiss Alps! She's studying biodiversity in high altitudes and what better way than being "there"? She heard the golden eagle and felt the virtual snow. It's almost like Mother Nature as the teacher.

Now here's where it gets even wilder. Remember when we used to raise our hands, waiting, sometimes desperately, to catch the teacher's eye for a question? In this classroom, an AI assistant patiently waits, ready to answer queries, provide additional resources or suggest a fun challenge to anyone feeling adventurous. All of this, mind you, gets recorded on a secure, tamper-proof Blockchain ledger. No more "my dog ate my homework" excuses; it's all there, transparent and immutable.

By blending AR, VR, AI and Blockchain, we're not just creating classrooms but crafting universes tailored to each learner. Universes where the shy kid in the back row can stand toe-to-toe with historical figures and where the loudest brainstorming doesn't come with the sound of chalk on a blackboard but with the excitement of endless possibilities.

REAL-WORLD APPLICATIONS

I've got some tales that'll knock your socks off. These aren't mere fantasies but real-life narratives from schools across the globe, where the magic of AR, VR, AI and Blockchain have all come together in harmony.

Greenwood High in Sunny California. Do you remember how we used to "travel" by just staring at flat maps in our textbooks? Well, not in Greenwood! Here, geography lessons involve VR headsets and a virtual globetrotting journey. One minute, students climb the Eiffel Tower; the next, they're snorkeling in the Great Barrier Reef. The coolest part? Their AI assistant quizzes them on the go, making learning an adventurous game.

Phoenix International School in Tokyo. Now, they've introduced a concept called the "AR Yearbook". Yep, you heard that right! Their yearbook is no longer just pages with photos; it springs to life when viewed through an AR app. Alumni tales, school plays, sports events—it's like living the memories again. All these videos and data are stored on a Blockchain, preserving them for eternity.

Berlin, the Urban Tech Academy has given a new meaning to 'personalized learning'. Using AI, they've developed a system recognizing each student's learning style and pace. Struggling in algebra? The AI recommends AR-based tutorials. Excelling in history? Dive into a VR experience of the French Revolution. It's like having a personal tutor, but way cooler!

The Everest Institute in Mumbai. They've gamified education using all these technologies. Imagine learning economics via a VR-based market simulation or understanding physics with AR models of the solar system. Every achievement and every test score is logged in a Blockchain-based digital report card.

ADDRESSING POTENTIAL BARRIERS

Cost: When I first got wind of these tech wonders, my jaw dropped looking at the price tags but where there is a will, there is a way. We already discussed shared resources and pooling funds but what about crowdfunding? Why not? It is easy for people to support great educational initiatives—after all, education shapes and affects us all, right? There's power in unity, friends!

Technical Complexities: Remember when Wi-Fi felt like rocket science? Or, when using a touch screen, a mind-bender? We got through that, and we'll get through this! Regular workshops and a dedicated tech support team can be our guiding stars.

Teacher Training: Between you and me, I've seen teachers break into a cold sweat hearing "Blockchain" or "virtual reality." We've all been there. Starting is always the hardest part. What we need are bite-sized training sessions. Think of them as mini-adventures into the tech world. Game-based learning for teachers? Heck yes! And for our educators who are a tad more hesitant, pairing them up with tech-savvy mentors could do wonders.

PREDICTING THE NEXT PHASE OF TECH INTEGRATION

Have you ever found yourself staring at classroom tech and thinking, "How did we ever manage without this?" Remember chalkboards? Projectors felt revolutionary back in the day! Now, with AI, AR, VR and what not, it feels like we're teaching aboard the Starship Enterprise. Here's the real question: where to go next?

Tech classrooms no longer have walls—I'm not advocating for tearing down walls (though, hey, open-air classrooms sound refreshing). I'm talking about virtual global classrooms, where a student in Tokyo could be group mates with someone from Toronto or Timbuktu. Connectivity will be the name of the game as we move from isolated classrooms to global learning hubs. Think "cloud classrooms" where knowledge floats seamlessly across geographies.

Picture vast, virtual libraries, accessible by a mere thought, thanks to brain-computer interfaces. Yep, you heard me.

Imagine attending a history class and downloading the entire timeline of the Renaissance right into your brain. Sounds far-fetched? Maybe now, but the winds are changing!

While we're on the subject of mind-blowing tech, let's spare a moment for holographic teachers. Imagine Albert Einstein breaking down his relativity theory right before you or Shakespeare reciting a sonnet in your own classroom. Wouldn't that be something?

But, as always, with great power comes great responsibility. We need to ensure that while we're tech-ing up our classrooms, we're also grounding our students in soft skills, ethics and empathy. After all, we don't want robots for students.

While these are just predictions, it's crucial for us—teachers, policymakers, students, parents and institutions—to prepare. To seek out the latest in tech, to understand it and most importantly, to ensure it's accessible to all. Education is, and always should be, an equalizer.

INFRASTRUCTURE AND TRAINING

We are all excited, right? Picture a teacher, perhaps like yourself, trying to set up a virtual global classroom. She's got the software and enthusiasm, but the school's Wi-Fi is slower than a snail in a traffic jam. The hardware is from the Stone Age and no one knows how to use the software. That is why infrastructure and training are paramount. It's like laying down the tracks for the Hogwarts Express and not having the magic to make the train appear.

First, we're talking about beefing up that physical foundation. Robust servers, speedy Wi-Fi and devices that don't scream "retro" (unless it's Throwback Thursday). Schools need to invest here. No shortcuts. Let's not forget accessible tech for students with special needs. Every child deserves to be on this tech train.

If a teacher is handed a gadget without the know-how, it's like giving me a lightsaber. Cool to look at, but dangerous if I don't know which end to hold. Regular workshops, hands-on sessions and maybe even a tech buddy system could work wonders. Balance is key.

THE ROLE OF EDUCATORS

It's safe to say that many of us love shiny gadgets and cutting edge tech, but this is not just about having the fanciest tools on board. It's about how we, the captains, navigate and use them to chart a course toward enlightening our young crew. Every piece of tech, whether an app or a device, is like a tool in our navigation kit. Use it wrong, and we might end up off course or stranded on a deserted island of confusion.

Enter the evolving role of educators. Gone are the days of "chalk and talk." Our methods must be as dynamic as the tech we use. Think of it as changing the sails with the direction of the wind. If there's an app that gamifies math, why not incorporate it? Virtual field trips? Heck yes, the idea is to weave tech seamlessly into our lessons, not just use it as a flashy add-on.

While we're on the topic of adapting, let's not forget the golden rule; not all tech fits all. Tailoring is crucial. What works for Jane might not work for John. We've got to be observant, flexible and a tad bit psychic (just kidding on that last one...or am I?).

Now, I will tell you a little secret: mistakes will happen. There'll be tech hiccups, moments of doubt and maybe even the occasional longing for our simple overhead projectors, but that's all part of the journey. It's how we learn, adapt and grow that matters.

CONCERNS IN A HIGH-TECH LEARNING ENVIRONMENT

DATA OWNERSHIP AND CONTROL

Imagine every assignment you hand out, every grade you receive, every single piece of data you generate, all zipping around the digital cosmos. It's like our own little constellations of information! But here's the million-dollar question: Who truly owns this data?

I remember the days when keeping student records meant rows upon rows of dusty filing cabinets. Fast forward to today, and voila! Everything's digital. But with this digital transformation, a wave of concerns crashes ashore. That's where our unsung heroes, Blockchain and AI, come into play.

With Blockchain, every bit of data has a clear, traceable origin and path. No more mysteries about where your data went or who accessed it. The best part? You hold the keys. It's empowering.

Meanwhile, AI (or, as I like to call it, our digital Sherlock) is continuously analyzing, filtering and optimizing—but don't fret! It's not here to take over the world. Its role? To serve us, ensure we make the most out of our data and keep it safe and sound.

Let's not just stop talking. Let's actionize. Educators, students and institutions must come together, explore and adopt these technologies. To truly own our data, we must understand, control, and, most importantly, protect it.

ADDRESSING BIASES

Ever tried on a pair of VR goggles or marveled at the wonders AR showcases? Perhaps you've seen AI in action, predicting, learning and evolving. It feels like stepping into the future but there's a flip side.

Imagine you're an avid traveler exploring a virtual city through your shiny new VR headset. Suddenly, you notice something. The representation of the local culture seems, well, off. Stereotypes? Check. Cultural insensitivities? Double check. Yikes! Now, you see where I'm coming from?

These amazing tech tools—AR, VR and AI—are like toddlers. They learn from the data we feed them. If that data is skewed, they'll paint a distorted image. We might inadvertently end up with tools mirroring biases we'd rather leave in the past. "Well, what can I do?" I hear you ask. Be curious and ask questions. Dive into the 'how' and 'why' of these technologies. Who's developing them? Whose stories and experiences are being included and most crucially, whose are left out? Inclusion isn't just a buzzword. It's our responsibility. As educators, policymakers, students and parents—we all play a pivotal role.

FUTURE GUIDELINES

We're bound to see a surge in the demand for ethical standards. With so many new gadgets and gizmos popping up, our dedication to protecting our students will be paramount. No one wants tech to be the Wild West anymore. The shift? Guidelines galore! Clear, concise and consistently updated. Because, let's face it, the rapid pace of tech waits for no one.

Remember when Pluto was considered a planet? Yep, science evolved and so did our understanding. Similarly, our teaching methods and tech ethics must play catch up. As AI gets smarter and VR worlds get vaster, we need guidelines that aren't just reactive but proactive, like being two steps ahead in a chess game with a cheeky computer.

I'm also willing to bet that collaboration will be huge! Tech companies and educational institutions will need to come together to create best practices and who better to ask, then the teachers?

KEY CHAPTER TAKEAWAYS

Integrated Learning Environments: The fusion of AR, VR, AI and Blockchain can craft tailored learning universes, making education more immersive and personalized.

The Evolving Role of Educators: With the influx of technology, educators must adapt, ensuring that tech enhances learning without overshadowing the human touch.

Data Ownership and Biases Prevention: As we embrace digital tools, safeguarding content ownership and promoting ethical tech use is paramount.

CHAPTER LEARNING ACTIVITIES

ACTIVITY 1: TECH IN YOUR CLASSROOM

Imagine you're an educator with unlimited access to AR, VR, AI and Blockchain. Design a lesson plan for a topic of your choice, integrating these technologies. How would you ensure that the lesson is both educational and engaging?

ACTIVITY 2: ETHICAL DEBATE

Organize a discussion or debate on the ethical implications of using AI in classrooms. Consider questions like: Should AI be allowed to grade student assignments? What biases might AI introduce, and how can they be mitigated?

ACTIVITY 3: FUTURE CLASSROOM VISUALIZATION

Using a VR or AR tool (or even just a simple drawing tool if unavailable), design what you believe a future classroom might look like. Think about the physical layout, available tools and how students might interact with the tech and each other.

18

SHAPING THE FUTURE: AI-ENHANCED IMMERSIVE EDUCATION

I recall a chat with my niece last summer. As we sat under the shade of the old oak tree in our backyard, she excitedly told me about her virtual trip to space. A trip that was part of her regular school curriculum! These immersive experiences were unheard of in my school days. Dare I say, learning is more memorable now, than twenty years ago.

I genuinely believe that this blend of tech will break down barriers. Barriers of geography, physical limitations and even socio-economic disparities. Everyone gets a front-row seat in this new age of learning. While there might be bumps along the way, the end game is worth it.

HOW THE LANDSCAPE MIGHT EVOLVE IN THE NEXT DECADE

First off, do you remember those bulky encyclopedias we once lugged around? Well, fast-forward ten years and our learning materials will be as fluid as water, constantly updating and adapting. I wouldn't be surprised if, by 2033, every student has an AI-powered digital mentor. Think of it as a personalized J.A.R.V.I.S from Iron Man but for education.

Swinging over to classrooms, and boy, are they in for a major makeover! Gone will be the rigid chair-desk setup. Instead, imagine flexible learning spaces, some real, some virtual, where students 'travel' to the African Sahara one day and the surface of Mars the next. The world (and beyond) will be our playground, thanks to AR and VR.

Speaking of play, remember dodgeball? It's great fun but not too educational. (Trust me, I've tried making a case for it!) The games of the future, though, will be different. Gamification will be the talk of the town. Kids mastering algebra while rescuing princesses or defending intergalactic bases?

Now, a little birdie told me that how we assess learning is also set for a revamp. Say goodbye to the sleepless nights before final exams. Continuous, real-time assessments will monitor students' progress, ensuring they grasp concepts at their own pace. It's not about cramming at the last minute but truly understanding and applying knowledge continuously..

But with advancements in brain-computer interfaces, who's to say we won't download knowledge Matrix-style? "I know Kung Fu," anyone? Okay, maybe that's a bit out there...for now.

On a more grounded note, the silos separating subjects will crumble. Interdisciplinary learning will be the norm. Art meets science. History dances with technology. Boundaries? What boundaries?

But while tech will play a colossal role, the human touch remains irreplaceable. Teachers will transition from information providers to guides, mentors and even co-learners. They'll foster creativity, critical thinking and emotional intelligence—aspects that machines, no matter how sophisticated, can't quite grasp.

If you've ever tried making sourdough bread (a lockdown favorite for many!), you know that patience, timing and the right ingredients are essential. Similarly, the future of education isn't just about flashy gadgets and sophisticated software. It's a delicate balance of tech, human connection, real-world challenges, and, most importantly, a relentless thirst for knowledge.

PREPARING STUDENTS FOR THE DIGITAL AGE

Whether you're a teacher weaving wisdom, a student embracing the data deluge, or a curious cat (and by a cat, I don't mean the furry kind, but you're welcome here, too!), the digital era is upon us.

When I was but a sprout, the height of technology in school was that fancy overhead projector with transparent slides. Today, it's a whole different ballgame. How do we prepare our vibrant young minds to survive and thrive in this brave new world?

First and foremost, we can't just throw tech at them and hope for the best. The digital age is more than just fancy gadgets. It's a mindset. It's about fostering a sense of curiosity, adaptability and resilience. The gadgets are just the cherry on top.

Speaking of resilience, remember dial-up internet? Oh, the agony of waiting for a page to load. Today, lightning-fast connectivity is the norm, but patience is still golden. In a world of instant gratifications, teaching students the art of patience, the value of delayed rewards and the beauty of perseverance is vital.

I'm not just talking about knowing how to tweet, snap or TikTok (though kudos if you've mastered the art!). It's about discerning real from fake, credible from incredible. Equipping students with the tools to analyze information critically is non-negotiable. After all, not everything on the internet wears its truth hat.

Now, you might wonder, "What about humanity?" While we're hurtling towards an increasingly digital world, human beings remain at the heart of it all. Machines may crunch numbers, but hearts and minds? That's our domain.

Imagine a cozy evening by the fireplace. The allure of a physical book, the turn of a page—some experiences are timeless. In the same breath, while we champion digital literacy, let's not forsake the analog. Balances are the key.

Much like we wouldn't send a child into the woods without a map, navigating the vastness of the digital forest requires its own set of survival skills. Privacy settings, secure passwords and understanding digital footprints are the compass, torch and Swiss Army knife of the online world.

BUILDING AN ECOSYSTEM OF INNOVATION

Picture a lush forest, teeming with life and interconnections, each tree and shrub coexisting harmoniously. That's the magic of an innovation ecosystem. Now, instead of trees, we've got tech gurus, educators, policymakers, dreamers and doers.

I remember those endearing moments of crafting DIY projects with whatever junk I could find around the house. Bits of wire, discarded batteries and old radio parts were a mesh of chaos and creativity. Fast forward to today and you might say our innovation landscape is a tad bit like my childhood tinkering, albeit on a much grander and more organized scale.

THE IMPORTANCE OF WORKING TOGETHER

Imagine tech giants, educators and policymakers sitting around a campfire, toasting marshmallows and plotting the future. Sounds dreamy, right? But it's not just about singalongs (though I'm always up for a good campfire tune). It brings together diverse minds and talents to craft a shared vision.

You see, tech companies are the wizards of our time. With a wave of their silicon wands, they conjure devices, software, and platforms that can dazzle and transform. Think about the first time you used a tablet in class or when AI made personalized learning a reality. That's some real sorcery right there but even wizards need guidance.

That's where our educators step in. They're the guardians of knowledge, the torchbearers who light the way. They understand the pulse of the classroom, the nuances of learning and the hopes and dreams of each student. Their invaluable insights can shape how technology unfolds in educational spaces.

Now, let's spice things up a tad and think of this collaboration as a rock band. The tech companies are the electric guitarists, educators are the soulful vocalists, but who's on the drums? That's right, the policymakers. They set the rhythm, create the framework and ensure everyone plays harmoniously. Without their strategic oversight, our concert of innovation might sound more like a raucous jam session.

Why, you ask, is this symphony of collaboration so darn important? Well, it's because the challenges of the 21st century are multi-faceted beasts. Tackling them requires a melding of minds, a fusion of expertise. Alone, each entity might make ripples but together, they can create waves of change.

I once watched a group of kids trying to solve a puzzle. Each had a piece, and they could have clashed, but instead, they communicated, collaborated and voila! The picture came to life. That's the essence of what we're talking about here.

The synergy between tech companies, educators and policymakers isn't just beneficial; it's essential. It's about crafting solutions that resonate and about ensuring that innovation benefits all, not just a select few. Whether it's building inclusive AI algorithms or designing tech solutions that bridge the educational divide, collaboration is the golden key.

ENSURING RESOURCES FOR INNOVATION

Let's say you're baking a pie. You've got a mouthwatering recipe, the passion, and the flair but hang on a second, where are the ingredients? Without them, that pie remains a compelling dream.

I've seen countless brilliant projects, bursting with potential, fizzling out just because they couldn't get the resources they sorely needed. It's like having a plant with no sunlight—all the potential but nowhere to go.

Remember those weekend garage sales? The ones where you'd find hidden gems amidst the knick-knacks? Funding innovative projects is somewhat like that. There's value everywhere, but one needs the keen eyes to spot it and the heart to invest in it.

Our educators, students and innovators are brimming with world-changing ideas. I've sat in classrooms and heard students pitch ideas that could make our Earth a better place. Yet, without the essential backing, these ideas could remain just that—ideas.

So, who's going to be our knight in shining armor? The venture capitalists? The government? Local communities? How about all of the above? Think of it as a collective responsibility. We're in this together. Weaving the dream tapestry of tomorrow's innovations demands everyone's involvement.

A wise old professor with a penchant for ornithology once told me that it's not just about the dollars and cents. It's also about mentorship, networks, expertise, and, sometimes, a friendly nudge in the right direction.

Now, for a fun twist, imagine a world where every young mind, every fiery educator and every starry-eyed dreamer has the backing they need, where innovation isn't stifled by the scarcity of resources but propelled by the abundance of belief and support.

STRATEGIES FOR ADDRESSING HESITATIONS/ CONCERNS

You know, that feeling when you're trying to introduce a new idea, and this metaphorical wall seems to pop up? Yep, we've all been there.

A while back, I was trying to get my Grandma to switch from her old flip phone to a smartphone. "Why would I need all those apps?" she'd ask, eyebrow raised. It took patience, understanding, and—most importantly—addressing her concerns head-on. Now that's just one grandma, imagine an entire institution filled with diverse individuals, each with their own set of concerns.

Changing your entire curriculum or adapting to a new technology can be daunting. It's like being handed a new recipe when you've been perfecting your grandmother's secret sauce for years. It's not about replacing but enhancing. We do not need to discard the old just yet, but how do we blend it with the new innovations on the horizon? Here are a few ways;

Through Open Dialogue: Remember those heart-to-hearts with close friends? When you truly listened and they felt heard? That's what we need. There is no judgment, just genuine conversation. Let's create spaces where educators and stakeholders can voice their concerns freely.

Through Education and Training: Knowledge is power or was it power is knowledge? Either way, the more we know, the less we fear. Host workshops and webinars and bring in experts. Let's show, not just tell.

Through Pilot Programs: A little experimentation never hurt anyone. Except for that one time with the blue hair dye, that's a story for another day. Testing new ideas in a controlled environment can help ease concerns and provide valuable feedback.

Through Showcasing Success: Remember that colleague who always has the latest gadget and can't stop raving about it? Become that person for innovation. Share success stories, the positive impacts and the transformations.

Through Collaboration: Two heads, they say, are better than one and when it comes to overcoming resistance, collaboration is key. Involve everyone in the decision-making process. After all, we're in this together, right?

BREAKTHROUGHS/MILESTONES IN IMMERSIVE EDUCATION

Do you know that exhilarating feeling when you finally nail that yoga pose after weeks of trying? Or when you finally figure out how to use the new coffee machine? That's how I feel about the steps we've taken towards immersive education. It's like a continuous series of "Eureka!" moments and I'm here for all of it.

I once overheard a student saying, "History class feels like a time-travel adventure!" Can you believe that? History is now mind-boggling and oh-so-cool! Who would have ever imagined?

When the first interactive whiteboard in my school, the excitement in the room was palpable. Everyone, including the usually reserved Principal, wanted a piece of that action. It's like when my neighbor got that new grill, and suddenly, everyone was a BBQ enthusiast. But I digress.

Another beautiful moment was when a teacher told me her students were excited about homework. Why? Augmented reality made math problems come to life. Imagine solving equations with dragons and unicorns! It's a math lover's fantasy that comes to life.

Now, let's not forget our fabulous educators. Teachers, lecturers, professors—the whole lot! They've rolled with every change, adapted to every new tool and provided the most engaging learning experiences. It's like watching a DJ seamlessly transitioning from one track to the next. Except in this case, the tracks are teaching methods and the dance floor is a classroom.

As for parents? They are coming on board too, because, well let's face it, most parents love seeing their children get excited about learning! I have heard of parents joining app review clubs and sharing app recommendations in chatgroups, so they can get the lowdown on educational tools. Just brilliant!

That's not all, let me shout out to the unsung heroes: the IT staff and the policymakers. The folks ensuring every firewall stands tall and every new method gets the green light. These successes aren't just milestones but beacons of hope, lighting the path toward a more inclusive, immersive and fun learning experience for all.

EMPOWERING STUDENTS TO THRIVE: LIFE-LONG LEARNING

Have you ever tried to use an outdated smartphone? The apps don't update and the system crashes. It's like trying to ride a tricycle in the Tour de France. Just like technology, the world of education is continuously evolving. We're not in the era of blackboards and inkwells anymore. We're in a dynamic digital age and our students must keep up, but how?

The answer is simpler than you might think: empower them to be lifelong learners. As teachers, we must instill the zeal to learn, not just for grades or diplomas, but for the sheer joy of it. Imagine diving into subjects headfirst, not because you have to, but because you WANT to. It's like how some people binge-watch shows on weekends; only here, we're binge-learning.

You know, when I was a kid, I believed that once you left school, the learning stopped. Boy, was I wrong? The world no longer operates like that (if it ever did). Let me share a little secret with you. Have you ever seen those people who are always curious, growing and ahead of the game? Yep, you guessed it. They're lifelong learners. They treat life like one big classroom and the digital age? It's their playground.

Gone are the days when you'd wait for the weekly magazine to get updates on your favorite topics. Now, the internet brings everything to our fingertips. It's like having a 24/7 buffet of knowledge and our lifelong learners are always hungry. But how do they do it? How do they not get overwhelmed with the sheer amount of information?

Firstly, they're selective. Like picking the juiciest apple from a tree, they know which sources of information are worth their time. Teachers, especially, have this uncanny ability to sift through the fluff. It's like having a built-in BS detector. Amazing, right?

Next, they're adaptable. Remember when the cassette tapes went out of fashion and we had to learn to use CDs? Then came MP3s and now streaming? It's the same with learning. Methods change, tools evolve but the thirst for knowledge? That remains constant.

Speaking of evolving tools, remember the first time you heard about online courses? I was skeptical. I thought, "Who'd want to learn from a screen?" Now, look around. Platforms like Coursera, Udemy and Khan Academy are booming. It's a testament to our adaptability.

But it's not just about adapting; it's about thriving in this digital landscape. To truly empower our students, we must cultivate a genuine love for learning. How, you ask? Well, the answer lies in engagement. Do you remember Mr. Henderson from 6th grade? The one who'd make history come alive with his theatrical antics? We need more Mr. Hendersons. Teachers who can make learning an adventure. Institutions that prioritize hands-on experience over rote memorization. Parents who encourage curiosity, even if it means answering a million "whys" a day.

For the students? Well, they are the real MVPs here. They must learn to juggle school, extracurricular and social lives and still find time to learn.

KEY COMPETENCIES IN A TECH-SOCIETY

Imagine, it's the year 2050. You're sitting in a self-driving car, dictating a message to your holographic personal assistant and your fridge at home is ordering groceries based on what's about to expire. If you had told me about this scenario 30 years ago, I'd have thought you were narrating a sci-fi movie plot but nowadays, anything is possible. Let's face it, the exact details of the future are unpredictable, and that's okay. One thing we can be sure of is that technology will play an even bigger role in it. So, here's the million-dollar question: What skills will our students need to survive and thrive in this tech-centric society?

Critical Thinking: Remember those puzzle games where you'd have to find your way out of a maze? That's how the future feels. With so much information at their fingertips, our students will need the ability to sift through the noise and get to the crux of any issue.

Adaptability: Ever tried using your grandparent's rotary phone? It's a trip! But just as we move from rotary phones to smartphones, the tools of the future will evolve, too. Students must be like chameleons, adapting to the ever-changing tech landscape.

Emotional Intelligence: Machines may be smart, but they don't have feelings (yet!). Building relationships, understanding emotions and navigating interpersonal dynamics will be more vital than ever.

Tech Literacy: It's not just about knowing how to use a device. It's about understanding the 'why' and the 'how' behind the tech. Our kiddos will need to know how to use technology responsibly and ethically.

Problem Solving: So, you've got a tech glitch. Do you panic or get to the root of the issue? Future challenges might not come with a manual, and our students will need to think on their feet, troubleshooting as they go.

I once tried to build a treehouse with zero carpentry skills. It was a disaster but it taught me one thing: always equip yourself with the right tools. For our students, these skills are their toolbox.

We are in a world where a single tweet can go viral in minutes and the line between the virtual and real has blurred. I remember when my biggest tech dilemma was choosing between two funky ringtones for my flip phone. Times have changed, haven't they? Now, we're not just talking about how fast our internet connection is or how many gigabytes our phones have; we're diving deep into ethics in technology.

Imagine a teenager, let's call her Mia. Mia's pretty tech-savvy—she can code and design her website and probably knows more about algorithms than I do about making a good cup of coffee. What happens when Mia reposts a meme on her social media without thinking twice about the implications it might have on someone's mental health?

This is where ethics comes into play. Being proficient in technology is like having a superpower; every superhero must learn the responsibility that comes with such power.

Understanding Digital Permanence: I'd sometimes doodle on the walls when I was young, only to be scolded later. But those doodles? They could be wiped clean. In the digital world, however, things aren't that simple. Every comment, like, or share leaves a mark. Our students need to realize the permanence of their online actions.

Empathy in the Virtual World: It's easy to forget there's a real person behind every online profile. We need to foster a sense of empathy in our students, reminding them that behind every pixelated photo is a living, breathing human with emotions and feelings.

Privacy and Security: I once had a friend who'd leave his front door unlocked because "he had nothing to hide." You can imagine my horror. Similarly, our kids might think it's okay to share every aspect of their lives online, but educating them about the possible repercussions is our job.

Ethical Use of Information: Imagine you're baking a cake and find a fabulous recipe online. Would you claim it as your own? Hopefully not! Similarly, in a digital age where information is at our fingertips, students must understand the value of giving credit where it's due.

ADAPTING TO NEW ROLES IN AN AI-ENHANCED LANDSCAPE

Remember Mr. Pearlman from grade school? The one with those gigantic glasses and that comforting aura? He was a master storyteller, weaving tales that transported us to distant lands. Now, imagine him with a bit of Tony Stark tech. Exciting!

So, what's our role in this AI-driven world? Are we educators becoming obsolete? Far from it! No matter how advanced AI gets, it'll never be able to replace the warmth of a comforting word or the joy of a teacher's proud smile. Our role is to be the heartbeat in a world of circuits and wires. We need to be adaptable and always learn. The day we stop learning is the day we stop teaching effectively.

Amidst a rapidly changing landscape, our core as educators remains untouched. We are the pillars, the constants in a rapidly evolving world. The tech might change, the tools might evolve, but the essence of teaching? That remains. Our role? To fuse the timeless art of teaching with the endless possibilities of technology.

KEY CHAPTER TAKEAWAYS

Preparing for a Digital Future: It is vital to equip students with the skills necessary to thrive in a technologically advanced society, including digital literacy and ethical usage of AI.

Future Educational Landscape: While the future maybe unpredictable, the evolution of education over the next decade must not overlook the importance of technology and lifelong learning

The Human Element in a Tech-Driven World: Teachers are the irreplaceable heartbeat of education in an increasingly digital education system.

CHAPTER LEARNING ACTIVITIES

ACTIVITY 1: THE FUTURE EDUCATOR ROLE-PLAY

Imagine you're an educator in a tech-enhanced classroom. Design a role-play emphasizing the importance of adaptability and lifelong learning for educators, parents and students.

ACTIVITY 2: ONLINE ETIQUETTE WORKSHOP

Organize a workshop to discuss the importance of online etiquette, include topics such as safeguarding against cyber bullying, respectful and appropriate communication and the importance of privacy.

ACTIVITY 3: DIGITAL-HUMAN DIAGRAM CHART

Using a VR or AR tool (or even just a simple diagram tool if unavailable), design a chart showing the distinct roles of technology (AI, AR, VR and Blockchain) and the human educator. Show the interconnections and distinctions of each.

CONCLUSION

We discussed almost every aspect of AI's limitless potential in education. From personalized learning pathways to global classrooms breaking the barriers of geography, AI presented a vision of an educational world filled with endless possibilities. As we saw, this is not just about algorithms and data; it's about connecting, understanding and enriching the human experience.

We also discussed the ever-important ethical considerations. AI, with its boundless potential, brings its set of dilemmas. It's not just about what we can achieve but what we should. The onus we discussed is creating an AI landscape that respects, understands and nurtures ethical boundaries, ensuring that our students are tech-savvy and ethically grounded.

We then discussed the role of educators. Far from rendering educators obsolete, AI promises to elevate their status, transforming them from mere information providers to guides, mentors and life coaches. In an AI-augmented reality, educators become the bridge between machine efficiency and human empathy.

In this ever-evolving, technology-driven era, the learning never truly stops. Both educators and students must understand the beauty of continuous growth, nurturing a mindset that doesn't see education as a phase but a lifelong journey.

If there's one word in my mind throughout this book, it's adaptability. You see, in our world of ever-advancing technology, staying still is akin to moving backward. The waves of change are relentless but exhilarating if you learn to ride them.

Imagine a surfer standing tall on the shore, watching as the tides of technology rise higher and higher. To some, these waves might seem daunting, even terrifying but to those who've cultivated the spirit of adaptability, these very waves are thrilling opportunities. Every rise, every fall, every twist and turn is a new chance to learn, grow and evolve.

The landscape of technology is in perpetual motion. While we've explored the magnificent horizons of AI in education in our chapters, it's crucial to remember that AI itself is just one chapter in the technological evolution. Who knows? Tomorrow, there might be another breakthrough or new revelation that could redefine our perceptions.

This is precisely why stakeholders—educators, policymakers, institutions and even you—must foster an agile mindset. The tools and techniques we discussed might evolve. New ethical dilemmas might emerge but the underlying essence remains the same—to be receptive, open and ever-ready to adapt.

It reminds me of a beautiful saying that goes, "Change is the only constant." The world of technology and education is constantly changing. It's about keeping pace, anticipating, innovating and sometimes even leading the charge.

Adaptability is more than just a skill; it's a philosophy. A mindset that doesn't see change as a challenge but as a cherished companion. After all, in the future, it's not the strongest or the most intelligent who will thrive but those most responsive to change. Let's envision a world where humanity and AI walk hand in hand, complementing each other in a harmonious duet. Throughout this book, one resounding theme has consistently echoed—AI, as wondrous as it is, serves as a tool but the real essence of education is irrevocably human.

Remember when AI hilariously misinterpreted a simple phrase or when we discussed the ethical dilemmas AI presents? Those moments of joy, introspection and even concern are quintessentially human. They're born from our ability to empathize, create, wonder and dream.

It's undeniable that AI can compute, analyze and even predict with accuracy, often beyond human capabilities. It's efficient, it's swift, and yes, it's transforming the way we teach and learn but can it inspire a student with a heartfelt story? Can it understand the subtle nuances of a learner's hesitation, unspoken fears or unbridled joy at understanding a challenging concept? That remains our domain.

The beauty of education isn't just in imparting knowledge but in the forging of connections—those magical moments when eyes light up with understanding, when barriers of confusion crumble and new vistas of imagination are unveiled. While AI can pave the way, our human touch, stories, compassion and creativity truly kindle the flame of learning.

Think of AI as a paintbrush, a magnificent tool that can help bring our visions to life. Yet, the artistry, the soul of the painting, is intrinsically human. It's born from our experiences, emotions and unique perspectives. Combining AI's precision with our innate human qualities creates a transformative synergy.

It would be oh-so-lovely if our journey with AI were just a bed of roses, wouldn't it? But let's keep it real. Every rose has its thorns and integrating AI into education is no different. While we've gleefully celebrated the myriad of opportunities AI offers, we shouldn't put on rose-tinted glasses regarding the challenges ahead.

Firstly, there's the big elephant in the room: equity. Ensuring that AI-driven educational tools are accessible to all, irrespective of geography or economic standing, is a monumental challenge. It's essential to recognize this disparity, or else we risk widening the educational gap further.

Then, there's the matter of data privacy. As educators and institutions rely more on AI tools, student data is floating around in the digital world and trust me, not all of it is locked up tight. Potential breaches, misuse of data or even unintentional sharing could have serious repercussions. Imagine a student's personal information, learning habits or academic challenges getting into the wrong hands.

Training and adaptation are also hurdles we can't overlook. Not every educator is tech-savvy. Remember Mr. Pearlman, our beloved history teacher who struggled to open an email attachment? Bless his heart. Expecting such educators to integrate sophisticated AI tools into their teaching seamlessly might be a tad optimistic. Comprehensive training, ongoing support and patience will be key.

And, of course, there's the looming question: Will AI replace educators? While we've established that the heart of education is irrevocably human, there are still fears and insecurities. We must navigate these concerns with sensitivity, emphasizing the complementary roles of AI and educators.

Amid all these challenges, it's easy to feel overwhelmed and question whether we're opening Pandora's Box but remember our shared optimism, our conviction? While the road ahead might be rocky, with collaboration, open dialogue and a sprinkle of innovation, we can turn these challenges into stepping stones for a brighter, more inclusive educational future.

To our dedicated teachers, thank you for your unwavering passion. As you venture into AI, stay curious. Continuous learning isn't just for your students; it's for you too! Attend workshops, engage with online communities and collaborate with peers. Always remember, while AI is a powerful tool, YOU are the artist. Use AI to enhance, not replace, your unique teaching style.

Dear students, the world is literally at your fingertips. Adapt AI as your ally. Use it to explore topics, solve complex problems or satisfy your curiosity but also take breaks, connect with nature and engage in face-to-face conversations. Balance is key.

Policymakers, you have the mammoth task of shaping the future of education. Prioritize equity. Ensure that every student, regardless of background, has equal access to AI-driven tools.

Invest in research, keep ethical considerations at the forefront and foster an environment where innovation thrives.

Educational Institutions, the age-old adage, 'Change is the only constant,' has never been truer. Upgrade infrastructure, invest in training for staff and actively seek feedback. Create an environment where teachers and students feel empowered to experiment and innovate.

And to everyone, a gentle reminder that while we're in the age of AI, the human touch remains irreplaceable. Value relationships, nurture empathy and cherish those 'Aha!' moments of discovery.

If there's one thing I've learned throughout my career, the thirst for knowledge is insatiable. When you think you've got it all figured out, along comes a curveball, and voilà, the learning starts anew. That's the essence of our journey in this AI-infused era.

For our hardworking educators, remember those early days when teaching felt like deciphering an alien language? Fast forward to today and you're mastering that language. As technology advances, you'll find yourselves as teachers and perpetual students, soaking in new methodologies, tools and perspectives. And trust me, there's a joy in that continuous evolution, a spark that keeps the flame of passion alive.

The world of AI opens doors to worlds previously considered science fiction. Virtual field trips to Mars, anyone? But remember, while the tools may change, the core remains the same: an insatiable curiosity. That drive, that quest for knowledge, will serve you well in school and life. You're not just preparing for a job or a career; you're gearing up for a lifetime of exploration.

A quick side note—lifelong learning isn't just about formal education. It's in the books you read, the conversations you have, the mistakes you make and the questions you ask. It's in the moments you pause to observe a butterfly or the nights you stay up watching documentaries about the mysteries of the cosmos.

In this AI-driven world, the lines between teacher and student blur. We're all learners, forever evolving, growing and adapting. At the end of the day, while machines may process data, the human heart processes dreams, hopes and aspirations. In this symphony of life, learning truly has no finale.

An old saying goes, "The journey of a thousand miles begins with a single step." When I set out to write this book, my dream was simple: to guide, enlighten, and maybe even inspire. Along the way, something magical happened. You breathed life into these pages. Your curiosity, questions and hopes for the future became the heart and soul of this work.

Looking forward, the horizon of education is both exciting and unpredictable. We stand at the cusp of a revolution, where classrooms stretch beyond four walls, teachers are not just mentors but co-learners and students become the architects of their destinies. With all its marvels, AI is but a catalyst, accelerating us into this brave new world.

But let's not forget that amidst the whirlwind of technology, the essence of education remains unchanged. It's still about igniting minds, touching hearts and shaping futures. Regardless of how advanced our tools become, they're a beacon that should always guide our path.

Before closing, let me extend my deepest gratitude. Thank you for reading this book and sharing your time, thoughts and dreams. It's been an honor and a privilege. As you close this book, I hope you carry forward not just the knowledge we've explored but the spirit of discovery, the joy of learning and the unwavering belief in the transformative power of education.

Till our paths cross again, keep dreaming, keep questioning, and above all, keep learning. The future is bright, and it's ours to shape with heartfelt thanks and warmest wishes for all your tomorrows.

INDEX

REFERENCES

10 Best AI Content Creation Tools in 2023 | ClickUp—https://clickup.com/blog/ai-content-creation-tools/

10 Best Examples Of VR And AR In Education—Forbes. https://www.forbes.com/sites/bernardmarr/2021/07/23/10-best-examples-of-vr-and-ar-in-education/

10 Educational AI Tools for Students in 2023 | ClickUp—https://clickup.com/blog/ai-tools-for-students/

8 Ways AI Is Used in Education—https://www.analyticsinsight.net/8-ways-ai-is-used-in-education/

A Brief History of AI in Education—https://drphilippahardman.substack.com/p/a-brief-history-of-ai-in-education

A Whole New World: Education Meets The Metaverse | Brookings—https://www.brookings.edu/articles/a-whole-new-world-education-meets-the-metaverse/

Abandon the Fear Around AI in School, says Saudi Education Minister: 'If You Wait To Check All The Boxes, You Lose The Opportunity'—https://www.aol.com/finance/abandon-fear-around-ai-school-154624936.html

AI For Content Creation: A Deep Dive—leewayhertz.com—https://www.leewayhertz.com/ai-for-content-creation/

AI in Education: A Practical Guide for Teachers and Young People—https://learnml.eu/docs/AI_in_Education.pdf

AI in Education: Change at the Speed of Learning—STEVEN DUGGAN—https://iite.unesco.org/wp-content/uploads/2020/11/Steven_Duggan_AI-in-Education_2020.pdf

AI In Education: Where We Are And What Happens Next—https://corp.oup.com/feature/ai-in-education-where-we-are-and-what-happens-next/

AI Technologies for Education: Recent Research & Future Directions—https://www.sciencedirect.com/science/article/pii/S2666920X21000199

AI Tools for Teachers | Edutopia—https://www.edutopia.org/article/7-ai-tools-that-help-teachers-work-more-efficiently

AI Tools in Teaching and Learning | Teaching Commons—https://teachingcommons.stanford.edu/news/ai-tools-teaching-and-learning

AI Will Transform Personalized Learning—https://www.fastcompany.com/90901403/four-ways-ai-will-transform-personalized-learning

AI Will Transform Teaching and Learning. Let's get it Right—https://hai.stanford.edu/news/ai-will-transform-teaching-and-learning-lets-get-it-right

AI-Assisted Special Education for Students with Exceptional Needs—https://www.researchgate.net/publication/374556232_AI-Assisted_Special_Education_for_Students_With_Exceptional_Needs

Artificial Intelligence and the Future of Teaching and Learning—https://warwick.ac.uk/fac/cross_fac/academy/activities/learningcircles/future-of-learning/ai__education_12-7-23.pdf

Artificial Intelligence and the Future of Teaching and Learning [PDF] — https://www2.ed.gov/documents/ai-report/ai-report.pdf

Artificial Intelligence In Education: Addressing Ethical Challenges—https://link.springer.com/article/10.1007/s43681-021-00096-7

Artificial Intelligence In Education | Unesco—https://www.unesco.org/en/digital-education/artificial-intelligence

Artificial Intelligence Innovation In Education: A Twenty-Year Data Analysis —https://www.sciencedirect.com/science/article/pii/S2096248720300369

Blockchain in Education—Office of Educational Technology—https://tech.ed.gov/blockchain/

Can AI Write a Good IEP? What Special Education Experts Say—https://www.edweek.org/teaching-learning/can-ai-write-a-good-iep-what-special-education-experts-say/2023/08

Comparing Student Achievement in Traditional Learning with a Blended Approach—https://www.ncbi.nlm.nih.gov/pmc/articles/PMC7308688/

Designing Environments Where Integrative Learning Can Flourish—https://www.centerforengagedlearning.org/designing-environments-where-integrative-learning-can-flourish/

Developing A Model For Ai Across The Curriculum: Transforming The Higher Education Landscape Via Innovation In Ai Literacy—https://www.sciencedirect.com/science/article/pii/S2666920X23000061

Eduaide: The AI Teaching Assistant—https://www.eduaide.ai/

Education Blockchain Initiative—American Council on Education—https://www.acenet.edu/Research-Insights/Pages/Education-Blockchain-Initiative.aspx

Ethical Principles For Artificial Intelligence In Education—https://link.springer.com/article/10.1007/s10639-022-11316-w

Experimenting With AI In The Classroom—Harvard Graduate School of Education—https://www.gse.harvard.edu/ideas/education-now/23/09/experimenting-ai-classroom

Exploring the Impacts of Generative AI on the Future of Teaching and Learning—https://cyber.harvard.edu/story/2023-06/impacts-generative-ai-teaching-learning

Generative AI And The Future Of Education: Ragnarök Or Reformation? A Paradoxical Perspective From Management Educators—https://www.sciencedirect.com/science/article/pii/S1472811723000289

Generative AI Has Disrupted Education. Here's How It Can Be Used For Good | World Economic Forum—https://www.weforum.org/agenda/2023/09/generative-ai-education-unesco/

Guidelines and Considerations for AI in Education—https://resources.finalsite.net/images/v1692133942/resanet/

ajemqpdmcryj2zc0nyaw/ai-guidance-document-august-023.pdf

How AI and Data Could Personalize Higher Education—https://hbr.org/2019/10/how-ai-and-data-could-personalize-higher-education

How AI Can Transform Education For Students And Teachers | World Economic Forum—https://www.weforum.org/agenda/2023/05/ai-accelerate-students-holistic-development-teaching-fulfilling

How Blockchain Is Used in Education | Maryville Online—https://online.maryville.edu/blog/blockchain-in-education/

How Could The Metaverse Impact Education? The World Economic Forum. https://www.weforum.org/agenda/2022/12/metaverse-impact-education-learning/

How Is AI Impacting Personalized Learning?—https://elearningindustry.com/how-is-ai-impacting-personalized-learning-key-applications-limitations-and-trends

How Robots Can Assist Students With Disabilities—https://www.nytimes.com/2022/03/29/technology/ai-robots-students-disabilities.html

Impact of Artificial Intelligence in Special Need Education to Promote Inclusive Pedagogy—https://www.researchgate.net/publication/342923570_Impact_of_Artificial_Intelligence_in_Special_Need_Education_to_Promote_Inclusive_Pedagogy

Improving Student Engagement in and with Digital Learning | OECD. https://www.oecd-ilibrary.org/sites/8a451974-en/

index.html?itemId=/content/component/8a451974-en

Kennesaw State Researcher Explores New Way To Use AI For Student Teacher—https://www.kennesaw.edu/news/stories/2023/researcher-explores-use-ai-student-teachers.php

New Era of Artificial Intelligence in Education: Towards a Sustainable Multifaceted Revolution—https://www.mdpi.com/2071-1050/15/16/12451/pdf

The Future of AI Education: Great Learning's Cutting-Edge AI Curriculum—https://www.msn.com/en-in/money/news/the-future-of-ai-education-great-learnings-cutting-edge-ai-curriculum/ar-AA1j95Wc

The Future of Artificial Intelligence in Special Education Technology—https://journals.sagepub.com/doi/10.1177/01626434231165977

The Future of Educational Content: AI's Transformative Role | Education Genius —https://educationgenius.io/blog/the-future-of-educational-content-ais-transformative-role/

The Intertwined Histories of Artificial Intelligence and Education — https://link.springer.com/article/10.1007/s40593-022-00313-2

The Impact of Artificial Intelligence on Learning, Teaching and Education: Policies for the Future — https://www.researchgate.net/publication/329544152_The_Impact_of_Artificial_Intelligence_on_Learning_Teaching_and_Education_Policies_for_the_Future

The Impact of Technology Integration on Student Learning

Outcomes: A Comparative Study — https://www.researchgate.net/publication/373266726_The_Impact_of_Technology_Integration_on_Student_Learning_Outcomes_A_Comparative_Study/fulltext/64e47c6040289f7a0faaf157/The-Impact-of-Technology-Integration-on-Student-Learning-Outcomes-A-Comparative-Study.pdf

The Promise of Immersive Learning: Augmented and Virtual Reality's Potential. https://itif.org/publications/2021/08/30/promise-immersive-learning-augmented-and-virtual-reality-potential/

Top Private School Appoints An AI Chatbot As Its New 'Headteacher' — https://www.dailymail.co.uk/news/article-12634065/Top-32-000-year-private-school-appoints-AI-chatbot-new-headteacher-UK-first.html?ito=native_share_article-nativemenubutton

Transforming Education: AI-Powered Personalized Learning Revolution. https://www.techopedia.com/transforming-education-ai-powered-personalized-learning-revolution

University of Warwick. Warwick International Higher Education Academy WIHEA — https://www2.ed.gov/documents/ai-report/ai-report.pdf

Using Technology to Improve Student Engagement | AACSB — https://www.aacsb.edu/insights/articles/2022/11/using-technology-to-improve-student-engagement

Using Technology to Support Student Engagement | Edutopia. https://www.edutopia.org/article/using-technology-support-student-engagement

ENDNOTES

[1] Independent School Parent (2023). How Will Artificial Intelligence Affect Education. Access at: https://www.independentschoolparent.com/school/artificial-intelligence-education/

[2] Sengupta, Trisha (2023). In A First, UK School Appoints AI In The Role of 'Principle Headteacher.' Hindustan Times. Accessed at: https://www.hindustantimes.com/trending/in-a-first-uk-school-appoints-ai-in-the-role-of-principal-headteacher-101697553952867.html#:~:text=As%20per%20reports%2C%20the%20organisation,human%20headteacher%20of%20the%20school.&text=Cottesmore%20School%20made%20this%20unique,the%20school%27s%20headmaster%20Tom%20Rogerson.

Printed in the USA
CPSIA information can be obtained
at www.ICGtesting.com
CBHW021630020724
10952CB00051B/20/J

9 781959 070290